"I love most in this fascinating book 'Some people are born wizards and other people have to work at it.' *The Physics of Miracles* is a handbook for becoming a wizard!"

C. NORMAN SHEALY, MD, PHD, professor of Energy Medicine, President Emeritus Holos University Graduate Seminary

"Richard Bartlett's done more to move the field of energy healing forward, and trained more healers in the past decade then anyone else has done since reiki was introduced to the Western world. His methods are repeatable, teachable, and great fun as is his dynamic voice and enigmatic personality that emanates through each page of this evolutionary, revolutionary, book."

DEBRA LYNNE KATZ, author of *You are Psychic, Extraordinary Psychic,* and *Freeing the Genie Within*

"Renowned healer Richard Bartlett drifts away from traditional science and ventures out into a discipline outside his academic expertise with a proposal of theories that go far beyond today's mainstream physics. This book invites you to look beyond personal concepts and biases. It will boggle your mind!"

KLAUS HEINEMANN, PHD (physics), author of *Expanding Perception* and coauthor of *The Orb Project*

"Bartlett's revolutionary healing techniques suggest that physics is missing something. When consciousness interacts with matter, surprising effects occur which defy current science. We used to call them miracles. His work awakens us to a new physics of parallel realities, not as an abstract theory but as a way to live fully and achieve our dreams. His work shows that we are truly co-creators of our experience."

DR. CLAUDE SWANSON, physicist and author of *The Synchronized Universe*

"What is this matrix that Dr. Bartlett is talking about? The best definition comes from the Sufi tradition: The matrix is that from which all things are continuously becoming. Dr. Bartlett's book gives us the rules that govern how things become, i.e., how we and everything in our world forms. These rules will make you laugh. The most important rule seems to be that there are no rules. The physicists will chuckle at this book. It will be good for their circulation."

JAMES L. OSCHMAN, PHD, 2008–09 president of International Society for the Study of Subtle Energies and Energy Medicine, author of *Energy Medicine* and *Energy Medicine in Therapeutics and Human Performance*

The Physics of
MIRACLES

Tapping into the Field of Consciousness Potential

Richard Bartlett, DC, ND

ATRIA PAPERBACK
New York London Toronto Sydney

BEYOND WORDS
Hillsboro, Oregon

ATRIA PAPERBACK
A Division of Simon & Schuster, Inc.
1230 Avenue of the Americas
New York, NY 10020

BEYOND WORDS
20827 N.W. Cornell Road, Suite 500
Hillsboro, Oregon 97124-9808
503-531-8700 / 503-531-8773 fax
www.beyondword.com

The information contained in this book is intended to be educational and not for diagnosis, prescription, or treatment of any health disorder whatsoever. This information should not replace consultation with a competent healthcare professional. The content of this book is intended to be used as an adjunct to a rational and responsible healthcare program prescribed by a professional healthcare practitioner. The author and publisher are in no way liable for any misuse of the material.

Managing editor: Lindsay S. Brown
Editor: Julie Clayton-Knowles
Copyeditor: Henry Covey
Design: Devon Smith
Composition: William H. Brunson Typography Services
Interior photos: Mina Bast

First Atria Books/Beyond Words trade paperback edition October 2010

ATRIA PAPERBACK and colophon are trademarks of Simon & Schuster, Inc.
Beyond Words Publishing is a division of Simon & Schuster, Inc.

For more information about special discounts for bulk purchases,
please contact Simon & Schuster Special Sales at 1-866-506-1949 or business@simonandschuster.com.

The Simon & Schuster Speakers Bureau can bring authors to your live event.
For more information or to book an event, contact the Simon & Schuster Speakers
Bureau at 1-866-248-3049 or visit our website at www.simonspeakers.com.

Manufactured in the United States of America

10 9 8 7 6

The Library of Congress has cataloged the hardcover edition as follows:

Bartlett, Richard,
    The physics of miracles : tapping into the field of consciousness potential /
    Richard Bartlett.
        p.   cm.
    Includes bibliographical references.
        1. Holistic medicine.   2. Mind and body.   3. Quantum theory.   4. Consciousness.
    I. Title.
    R733.B2754   2009
    613—dc22
                                                                        2009022511

ISBN: 978-1-58270-247-6 (hc)
ISBN: 978-1-58270-249-0 (pbk)
ISBN: 978-1-4391-5816-6 (ebook)

The corporate mission of Beyond Words Publishing, Inc.: *Inspire to Integrity*

To Dara Louise Bartlett,
my little one, who has taught me to know more,
love deeper, and to be expansive joy.

I also want to dedicate this work to
Lt. Col. Tom Bearden, PhD,
whose brilliant concepts concerning scalar electromagnetics
and its applications have inspired me and countless others.
He is a tireless warrior for truth and a defender
of this nation's sacred freedoms.

CONTENTS

ACKNOWLEDGMENTS

This book would not have been possible without a number of key players. First and foremost, extra-special thanks are owed to Melissa Joy Jonsson for her loving care and overseeing of every aspect of the transcription and editing of this book. To Cynthia Bartlett for having the initial vision to contact Beyond Words and establish a working relationship and for overseeing with loving eye and intent Matrix Energetics from its inception. To my publisher, Cynthia Black, for taking a chance on an unknown author and for believing in and supporting me each step of the way. I would also like to acknowledge the unceasing efforts of Daphne Hoge, who understands the principle of keeping "the talent" happy. To Lindsay Brown and Julie Knowles, my principal editors, whom I gleefully tortured with minutiae and details, I am especially grateful. To Lisa Braun Dubbels, my publicist who tirelessly works on my behalf.

I especially want to thank my children: Victor, who has done some of the principal artwork for Matrix Energetics; Justice, who has taught onstage with me and embodies the principles in this book; and Nate, who is always there for me, providing love, support, and everything else I might need. Finally, I want to thank the staff and students of Matrix Energetics, including Ben, Alyssa, CeCe, Brandon, Carol, Karen, William, and Rebekah.

INTRODUCTION

Everyone would love to experience a miracle, but doesn't it seem that there's never a miracle when you *really* need one?

There is a very tangible and real energy deeply embedded in this book. Like one of Harry Potter's magical texts, this book possesses a field of enormous potential. Cradled lovingly within these pages is an incredibly powerful energetic template. You too can use this consciousness technology, as thousands of others already have, to create "healing" for yourself, your family, and perhaps even the environment you live in.

What this book will show you is how what we call "miracles" in this day and age are actually indications of things already known to exist but not yet "proven." However, it has been said before that "absence of evidence is not evidence of absence." When we change our consciousness around what is possible, rather than being limited by a reality construct dominated by what isn't possible, we discover that we are actually able to employ quantum energies and principles in our day-to-day lives in unexpected, and fun—and miraculous—ways.

A miracle transcends the limitations of most of our current scientific models, but this does not mean that there isn't a science that can explain miracles, if we really need to have explanations. Some might say that some of the ideas in this book are "fringe science." Well, I would rather be out on the fringe with the Angels and miracles and the other wondrous things science can't adequately "explain." That's where the magic lives. And what if there were a way that you could increase the incidence of miracles in your life? *"Would that be worth something to you, laddie?"* (as *Star Trek's* chief engineer, Mr. Scott, might say).

Keep in mind that all matter is comprised of energy, or fields of energy: humans, animals, stardust, trees, chairs ... all form is made up of energy at a quantum level, and this is where the science of our so-called miracles occurs. Rupert Sheldrake is a biologist and prolific author who introduced mainstream audiences to the idea of the *morphic field* (see also *morphic resonance* and *morphic unit* in the Matrix Glossary):

> *A field within and around a* morphic unit, *which organizes its characteristic structure and pattern of activity. Morphic fields underlie the form and behavior of holons or morphic units at all levels of complexity. The term* morphic field *includes morphogenetic, behavioral, social, cultural, and mental fields. Morphic fields are shaped and stabilized by* morphic resonance *from previous similar morphic units, which were under the influence of fields of the same kind. They consequently contain a kind of cumulative memory and tend to become increasingly habitual.*[1]

TAPPING INTO THE MORPHIC FIELD OF THE MATRIX

Matrix Energetics has a huge morphic field that allows you, with minimal effort, to step into a unified field of consciousness. Thousands of people, some traveling from far corners of the world, have attended the Matrix Energetics seminars to date and have read my first book, *Matrix Energetics: The Science and Art of Transformation.*[2] Everyone who has participated in some way with this Matrix has contributed a certain momentum of energy to Matrix Energetics' collective morphic field, or grid.

This powerful group dynamic allows you to amplify each other's desirable momentum and abilities, in service of the collective good. It can only benefit you to plug in. Since it already exists, you can both re-create and access this field anywhere, at any time. When you tap into the existing morphic field of Matrix Energetics, you are stepping into what amounts to a highly organized and powerful *consciousness technology*. This technology contains awesome "software" and features great technical support "in the field."

You hold within your hands a morphic unit of the actual energy field of what I call *Matrix Energetics*. There is a great undercurrent of power contained within the pages of this book. It is not necessary to understand any of the scientific language within to benefit from the power it holds. This information can change your life in strange and wonderful ways! Magic and miracles are the same in any language, and they can and will speak directly to your heart.

This book is intended to be deeply absorbed into your unconscious awareness. To that end, it speaks to the reader in many different ways and on a number of levels. Go easy on yourself if you don't understand an unfamiliar concept or if the subject material at times seems slightly beyond your conscious grasp. When you are reading this book, allow yourself to temporarily relinquish what you *think* you know. You can even skip the more scientific-looking material if it's not your cup of tea. The energy that is deliberately encoded within this book is available to you whether you understand it or not.

Many of the stories and concepts in the book are humorous; sometimes I am downright and unabashedly silly! You can take the concepts I share with you and use them as you see fit. Most important is to have fun with this material and watch how your life will change.

A young man shared an incredible story with me at one of my seminars. This man had been illiterate all his life, and yet he had developed his own unique way of reading. He would go to the library, check out a book that "looked interesting," and take it home. Unable to read it in a conventional manner, he would put it under his pillow at night and literally sleep on the information. The next day when he awoke, his mind would inexplicably "know" what was in the book.

And perhaps equally amazing, the greater the number of people who had checked out the book previously, the greater access he had to the knowledge within the book. So if fewer people had read it, the details the man received were considerably sketchier in the scope and quality of the information he would receive. I believe that he was able to perform this amazing feat because of the existence of morphic fields that Sheldrake talks about.

This powerful group dynamic allows you to amplify each other's desirable momentums and abilities in service to the collective good of all concerned. It is of benefit to you to plug in. You can re-create this grid anywhere, at any time. When you tap into the preexisting morphic field of Matrix Energetics, you are stepping into what amounts to a highly organized and powerful consciousness technology.

A QUANTUM MIRACLE STORY

I really appreciate the point Dr. Bartlett makes that we should not think that we are doing any "healing." I have a strong belief in God, and I often feel conflicted between that belief and my training with the traditional model of Western medicine, which is characterized by its emphasis of science over belief in spiritual values. I do not want to conflict with God, so when Dr. Bartlett says to stop thinking and that you are the healer, this really makes sense to me. Really, what we are doing is allowing for God, or God's grace, to do the work. This allows me to do my work comfortably and humbly.

As a medical doctor, I have two specialties: nuclear medicine (quantum physics) and pediatrics. I retired approximately one year ago and am now studying for a master's in holistic health. Without ever having attended a Matrix Energetics seminar, I have already been able to help two different people and in three situations. All I did was watch a video on YouTube of Dr. Bartlett doing the Two-Point, and along with God's help, these people have gotten better, in seemingly miraculous ways.

The first time I tried Matrix Energetics was when a gentleman called me up from church and said he could not urinate. He had blood in his urine. He wanted medical advice, so I told him to go to the nearest hospital. He could not afford it. The next day, I felt bad that I was not very compassionate and that I didn't help him. So I decided to track him down and try to help him. I wanted to heal him, even though I know that it is really God who heals.

Eventually I found him in the community and asked him to come to the hospital for a blessing. When he arrived, I gave him the spiritual ministering. Then I decided to do a simple Two-Point: one in front and one in back. In the middle of this two-point process, the gentleman started urinating. To you this means nothing; to him it was miraculous. Within a day, all his urine tests were clear and he was out of the hospital the very next day.

The next person was a classmate of mine in an "energy healing class" who had a sinus problem. She had had surgery on her sinuses, but it did not help. She was always congested during class. During our breaks, we were supposed to practice quantum-touch technique. I thought what I was learning in the class would be too slow for her needs, so I decided to try Matrix Energetics. I created an imaginary holographic image of the sinuses and "noticed" where there were blockages, and then I just fixed them. Her sinus cavity cleared "just like that" in less than one minute!

Later on, she had another problem: she had torn a muscle in her leg, and although she had been trying acupuncture for several months, she was still in a lot of pain. She asked me to help. At first I tried to use the healing modality I was being taught, but it didn't appear to do anything. Then I tried the Two-Point from Matrix Energetics.

I did the basic Two-Point without knowing anything about it other than what I had seen on YouTube. I felt what seemed like the muscle slapping back. In my experience, it is very difficult to heal something like this when it is anatomically out of arrangement. But when I saw what Dr. Bartlett was doing with bones and muscles using two-pointing, I thought to myself, "Wow. This is something way beyond what we understand, but I think it can be useful." After I did the Two-Point on my classmate, she went running— and her pain was gone.

—D.S., *university professor,*
doctor of nuclear medicine and pediatrics

I personally like a book that challenges me. That's when I know that I am really meeting and (hopefully) transcending my personal boundaries and limitations. The information in this book is quite possibly unlike anything you have encountered before. Or perhaps it will serve as validation for what you already know in your heart but haven't been able to put into words or action. In any event, you might like to open the book at random to a page and read a little bit, and then "sit with the energy" and let it seep in slowly and gently. The magic contained within will work for you even if you don't believe that it possibly could!

With this book, you will plug directly into the field of dreams and miracles that many thousands of people have already built for you. You get to experience all the benefits with none of the hassles. It is easier and more fun than you can possibly imagine to create quantum miracles in your life.

1

IT'S A BIRD! IT'S A PLANE! IT'S...SUPERMAN?

My life, which is never what one could call entirely normal, turned completely bizarre in February 1997 when I saw Superman as a three-dimensional hologram while I was engaged in my busy chiropractic practice. Yes, it probably was a hallucination, but this particular one enabled me to heal the eyesight of a child instantly!

If you have read my first book, you are familiar with the story of how Matrix Energetics first came into being. Essentially, I was confronted with a young girl with amblyopia, commonly known as lazy eye, and I did not know what to do. Most of the tricks in my medical bag—the one with crystals and other strange and wonderful things in it—would be of no help in this circumstance.

I had driven up from Seattle to Livingston, Montana, where I was practicing as a chiropractor for the weekend in the office I owned. Having slept very little and worked all day, the veil between fantasy and reality, or alternate dimensions of possibility, was thinner than usual for me. (You can read the whole story in chapter one of my first book.) I still don't know what *really* happened, but the myth surrounding those events will do just fine. When confronted with something I did not know how to help or to heal, my subconscious mind offered to me the image of George Reeves as Superman, a powerful and useful archetype from my childhood.

One of Superman's many abilities is the power of x-ray vision, a very useful tool for a doctor to possess. I actually saw, through the projected image of Superman's x-ray ability, what was wrong with my little patient. I have never been one to question when the weird or wonderful shows up to offer me assistance. I acted immediately on the information I was presented with. Something new and amazing occurred, and the little girl was healed in that very moment.

Suffice it to say that it was entirely reasonable for me to project a holographic image of Superman from my subconscious outside of myself in order to solve the problem that was presented to me. I have always wanted to possess x-ray vision or clairvoyant sight. *Being Superman is a way to be able to see clairvoyantly without all the responsibilities of being a doctor who has x-ray vision!* I mean, imagine what it would be like if you saw clairvoyantly all the time. Perhaps you would have to wear special dark glasses like the *X-Men* character Cyclops. You could take off the shades when needing to "see," put them back on when the job was done, and then go on living a relatively normal existence.

While in chiropractic school in the 1980s, I saw a movie called *X: The Man with the X-Ray Eyes,* starring Ray Milland. In this movie, Ray played Dr. James Xavier, a surgeon who injected himself with a serum that allowed him to see right into people's bodies. This is a rather good ability for a surgeon to possess, you might think. In the movie, the hero saw that the chief surgeon was going to perform a life-threatening surgery on a young girl, based upon a wrong and potentially lethal diagnosis. Ray tried to convince the surgeon not to operate, but the girl was operated on anyway. She died, just as predicted. Ray's character was blamed for the other surgeon's screwup, and he became wanted for murder.

Taking refuge, he joined a circus carnival as a fortune-teller. There was a particularly moving scene in which a superficially minded woman wanted to hear about all the good things that were going to happen to her. Fortune-telling is supposed to be for entertainment purposes, right? Instead, Ray, looking at the malignant tumors growing unchecked in her body, told her the truth as he saw it: "You are going to die young, soon, and in horrible pain."

Everywhere the fugitive doctor looked, he saw disease and suffering. Finally, unable to stand it any longer, he ripped his eyes out of his sockets. (Well, it was a "Hammer Horror" film, so there had to be an ending like that. Now perhaps you understand why I subconsciously chose to let Superman be the go-between for my clairvoyant vision! See, it all makes sense now: let Superman do it or rip your eyes out in horror. Hmmm . . . that's a hard choice. Let me think about that, and I'll get back to you!)

The very next day after the "Superman Event," as the myth has come to be called, there was something vastly different about my energy. Suddenly, the act of lightly touching a patient with focused intent created dramatic, often startling changes. Bones would realign themselves, patterns of chronic pain would disappear, often with only one brief session, and scoliosis curvatures would realign right before my eyes.

Word traveled quickly among the members of the spiritual community that I was part of at the time. Everyone was talking about Dr. Bartlett and his miracle hands, and everyone wanted an appointment with me NOW! That first day after the Superman Event, there were so many people clamoring to see me that I worked until well after midnight.

Finally, when the last patient had been seen, I collapsed gratefully in my bed for a few hours of well-deserved rest. Unfortunately, the next day was Monday, and I was expected at school. With the heavy class schedule I was carrying, I couldn't afford to miss a single day. So a few hours later, at 4:30 AM, it was time to sing, "Hi ho, hi ho, it's off to school I go!"

One cannot burn the candle at both ends indefinitely and expect to get away with it. I was no exception. Upon returning to Seattle, I promptly contracted a nasty flu that incapacitated me. Perhaps due to my depleted reserves, the flu was only the first stop on my forced house rest. Within a week, my flu had become walking pneumonia. It was two weeks before I could even raise my head from my pillow, much less attend classes. At the end of my forced convalescence, my friend Debbie visited me at home and provided a much-needed massage.

Recovering somewhat under her gentle ministrations, I roused my energy enough to reach out and touch her. She looked shocked. "Richard, what is that incredible energy?" she asked.

"You can feel it?" I asked her. "Oh, thank God. I've been so weak that I didn't even know if it was still there!"

"It's the most incredible thing I have ever felt. What is it?" she asked again. So I told her the fantastic story about the Superman Event, careful to leave nothing out. She was, understandably, amazed.

"If you ever figure out how to teach this, if it can be taught, then I want it. You must teach me!" she insisted. I had no idea what would later transpire as a result of this conversation. At the time, Matrix Energetics was not even a gleam in my eye, but I promised that if it could be taught, I would happily teach her.

As time went on, this phenomenon seemed to grow stronger. Sometimes it revealed new and marvelous outcomes each week. Chronic conditions began to change even though they were often not even a conscious focus for the treatment. In addition, people began to report that their emotional states, belief patterns—their very lives—were mysteriously being transformed. Better still, these changes appeared to continue over time. My practice, always emotionally satisfying, became a profound and deeply moving day-to-day experience.

MIRACLE ON THE MOUNTAIN

Four months after the Superman Event, it was lightly snowing in Bozeman, Montana, and I was driving my daughter home to Big Sky, where she lived in a trailer home with several friends. Although it was the middle of June, and snow was somewhat unusual for that time of year, it was not all that unexpected. I carefully drove up the winding mountain road in my 1966 GTO.

I hadn't seen my daughter Justice very often since I had moved to Seattle to attend Bastyr's Naturopathic program. You have to be *some kind of crazy* to go back to medical school at the age of forty-two. After all, I had already been working as a chiropractor since 1987. During my first year in practice, I barely made enough money to feed my family. Unfortunately, my first marriage didn't survive the stress of those lean early years. Now ten years later, I was making a good, regular income. What insanity led me to think that I could—or should—go back to school and do it all over again at this age?

I was taking thirty-one credit hours a semester. What that meant was eating, sleeping, and living for school. Although I attended medical school full-time in Seattle, every other weekend I drove for thirteen hours to Montana (where I still retained an active license) so I could practice Chiropractic and put food on the table.

The only reason I got to see Justice at all was because I didn't have a chiropractic license in the state of Washington. In order to arrange a little face time with Justice, she had agreed to play secretary for my still-thriving practice whenever I was in Montana. We had just finished a grueling ten-hour workday, and I was driving her home.

The trip was uneventful and we were mostly content listening to the powerful roar of my Pontiac's 389 c.i. engine as we carefully wound our way up the icy mountain road. Safely arriving at my daughter's doorstep, I gave her a hug and told her I would pick her up first thing in the morning. As I drove off to return to my sleeping quarters in Livingston, I had no idea how my life was about to be turned topsy-turvy again!

Driving slowly back down the winding road, I suddenly saw something in the road up ahead. Was it really there? Or was what I was seeing just a hallucination? (When you've had an experience where you have seen Superman in a clinical setting, reality is pretty much up for grabs.) Slowing down, I looked more closely. Yes, there was definitely someone or something standing in the middle of the road. I pulled over and stopped my car.

Suddenly I could see clearly, with my mind's eye, a vision of a man wearing a brightly colored turban standing in front of me, his right arm upraised in greeting. Not wasting a moment on pleasantries, he addressed me in a powerful, commanding tone, saying, "I will let you have what you desire if you promise never to raise your hand in order to harm any part of life." I immediately nodded my agreement to his terms. With a gentle bow of his head, he looked me in the eye and commanded, "So let it begin!"

In the next moment, I heard a whooshing sound and felt as though the top of my head had been pried open and its contents were gushing out of my skull and into the sky. In my mind's eye, I saw and felt something huge descending from directly overhead.

Have you seen the movie *Close Encounters of the Third Kind*? Do you remember the scene where the giant, multistoried spaceship descends onto the landing platform on top of the mountain? That is the closest analogy I can give you to what I experienced in the next several minutes. Perhaps it will help you to imagine a giant, multicolored spinning top, and you will have some idea of what I felt.

As this thought-form construct opened, I began to have rapid images and scenes flash into my awareness. A massive "spiritual software program" began to download into my brain. Even though this program continued for more than a few minutes, I quickly became acclimated to what was occurring and regained the normal use of my bodily functions. The Master appeared to me again and told me to touch my left rib cage with my right hand. Obediently, I did what was asked of me. The moment my fingers touched my chest wall, there was a loud crack and my entire ribcage physically shifted into a new position.

You see, I was born with an extra lumbar vertebra. To compensate, my rib cage had twisted in such a manner that it had always protruded uncomfortably on my left side. This distortion was so prominent that when I was young, if I lay on my left side, I could feel the thrumming of my heart beating against my chest. I could not sleep on my left side because the vibrations of my heart against my rib cage would keep me awake.

However, as soon as I touched my left side as bidden by the Master, I was healed and noticed my rib cage reformed in a new and much more comfortable position. Tears of gratitude flowed down my cheeks; I could scarcely believe what had just occurred. In the blink of an eye, I was healed of a lifelong uncomfortable condition! The Master nodded serenely once again and disappeared.

When I had collected my emotions and wits sufficiently, I continued the thirty-minute drive back to Livingston. As I was driving, two Angels (that is the only word I can use to name them) provided me with a running discourse in my head about what the new spiritual software was meant to do. This conversation went on for the next five hours and was one of the most uplifting events in my entire life!

I have always regretted that I did not write down any of the things I was shown that night as there was much instruction about the new energies that I was carrying and how to use them for healing. Some of the discussion was quite technical and involved sacred geometry, the activation of the chakras in my hands, and so much more. As the angelic download drew to a close five hours later, a beautiful feminine being said to me, "This event was scheduled for your birthday [in May], but we needed a number of uninterrupted hours alone with you in order for this process to be completed. Happy Birthday!"

2

FROM THE BEGINNING: HOW MATRIX ENERGETICS STARTED

It was a bright summer day in Seattle two years later. On that particular morning, I was waiting in the Bastyr clinic for my next patient to show up when there was a knock at the door. When I opened it, there stood an unassuming blond man.

MARK ENTERS MY LIFE

"You're Dr. Bartlett, aren't you? I have heard a lot about you, and I am really glad to meet you. I have studied all these different healing techniques and wondered what you would suggest I study next," he said, handing me a piece of paper containing a long list of techniques.

Looking at the paper, I realized that I had studied everything on his list and had mastered many of the techniques.

"Who you think I should study with next?" he asked.

"Since I know all these techniques and can probably even teach you some shortcuts, I suggest that you study with me," I replied.

"All right, I'll do that!" he responded enthusiastically. Little did we know at the time that this was the beginning of a lifelong friendship.

Over the next four years, I taught Mark Dunn everything I had learned in my twelve years of unconventional and downright "weird" medical practice. The one thing he couldn't get, much to his increasing frustration, was how to do what I did with my hands—what he called that "energy thing."

You see, Mark had set up his reality in such a way that he misunderstood what I was doing. He thought I was "doing something" with my hands and there was some transfer of energy. Because Mark continually questioned my process, I began to read books about quantum physics in an effort to understand and explain just what was happening with my "energy thing." It was within the more esoteric aspects of *quantum field theory* that I begin to form the foundational principles that underlie what is now called Matrix Energetics.

To Mark's chagrin, my second student learned the basic principles behind what I knew in about four hours. Today, Matrix Energetics can be taught to anyone in a single weekend seminar.

The first Matrix Energetics seminar was taught in September 2003 to a total of twenty-seven people. Since then, the Matrix Energetics experience has become an international phenomenon. You do not even have to attend a seminar to learn Matrix Energetics (although they are a lot of fun). Many people have learned Matrix Energetics and have experienced miraculous results in their lives from simply reading my first book or watching me demonstrate the "techniques" on YouTube. (If you have not yet read my first book, you might benefit from doing so.)

This second book will take you far deeper into the Matrix and accurately reflects my most up-to-date theory and information. I have read more than a hundred texts about quantum physics in order to understand the little bit I now know. Mostly, I learned how much I didn't know—and would perhaps never understand. Even the famous physicist Richard Feynman knew that if you think you understand quantum physics, then you have not understood what the theory says.

Mathematics and physics are unconscious symbolic languages. They are meant to stimulate higher brain function—not the hired hand left brain's higher analytical functions, but, instead, the right-brain awareness of symbolism and interconnectedness. The problem is that we have made science a *closed system*. The equations we devise to explain the unexplored or unknown have to make "sense" to our left brain's way of thinking and perceiving. If it does not make sense and fit into our closed loops of reality, then we will just chop pieces off until it does. You cannot imagine how many times this has happened in the history of physics.

This book will talk about the technology and science of the forbidden *aether physics*. When you read more than 140 science and physics-related books in a few months, you begin to notice some glaring discrepancies in the classical or standard physics model. You begin to suspect that maybe there are a few pieces missing here and there.

The problem with the classical model of physics is that the x factor of consciousness is not taken into account. Without the acknowledgment of the role consciousness plays in our experiments performed in life's laboratory, what you get are the unconscious expectations of the observers performing the experiments. The standard model in physics is a closed system. In a closed system, the effects of consciousness are excluded. In such a system, there is no room for miracles.

MATRIX ENERGETICS AS A USEFUL MYTHOLOGY

Through Matrix Energetics and within the pages of this book, I am going to share with you a powerful and useful mythology. Keep in mind that anything you or I embrace as truth is probably somewhat of a distortion. We put our central concepts and cherished ideals in play until something better comes along. Consider some of the things I share with you as placeholders until something better arrives at your doorstep.

Always champion your highest truth for the expression of your greatest good. My spiritual teacher, Elizabeth Clare Prophet, used to tell me that what she gives to the world is her best effort. She emphasized that if she ever sees anything better that rings more true for her, she would be there in a heartbeat. Be a servant to your heart's inner truth, and your life in turn will bow to you as its Master.

The less you know about physics and math, the more you are probably going to embrace the ideas in this book. From a purely scientific perspective, if you are a physicist, you will see this material quite differently. Please don't let that stop you from considering some of these concepts.

Go ahead and help yourself to an extra portion of nonsense, if that's how you perceive it. It could just be good for you. At least it doesn't taste like brussels sprouts! Some things, even though they might be good for you, can just leave a bad taste in your mouth.

It really doesn't matter if you believe me or not; these ideas can still change your life.

MATRIX ENERGETICS: WHAT IT IS AND WHAT IT ISN'T

There are no hard and fast set rules that say, "This is Matrix Energetics." Matrix Energetics doesn't exist as such; I made it up. Well, I made up the name and the tools to access and understand it, but there are also underlying principles at play, which I didn't make up. However, very few people will admit this about their so-called techniques or methods. I made it up, but I did not invent the science that I present within these pages.

I have attempted for this last half of my life to serve and heal the sick. I have always desired to provide comfort where it is needed. Above all, I have tried to stay curious, humble, and ever questioning. I can now serve many more people than I could ever reach and help in private practice. Through the Matrix Energetics books and seminars, I have been able to positively impact the lives of many thousands of individuals throughout our planet. I hope that this book can be a stepping-stone for you to new levels of possibility and mastery.

3

BUILD TRUST IN THE UNKNOWN BY ASKING OPEN-ENDED QUESTIONS

By cultivating the habit of asking powerful, mind-altering questions, you are training your right brain to respond to the signals from your subconscious.

The reason I ask questions is not because I know anything. I do not. I ask a question because I am curious about whatever engages the lens of my attention: "What is useful about what I am noticing in the moment?"

To get "in the moment," ask a question, step back, and observe any internal or sensory response. When I say "observe," I am using visual language. However, anything you notice from the domain of your five senses after asking the question is potentially useful and possibly life altering. At this point, all I'm asking you to do is to ask the question and then suspend any preconceived thoughts around it—and notice anything that is different from what you were noticing before. I know this sounds somewhat cryptic right now, but it will all make more sense as we go along.

Persistence and belief are the keys. Believe in yourself. You have the right to guidance, love, and assistance from the realms of Spirit. Whatever question you ask sends your mind on a search for the answer to your query. Your brain, at certain levels of activity, functions just like a machine. Whatever input you provide will be received in the language of the realm with which you frame your request. If you expect to get visions or dreams from beyond the veil of conscious reasoning, you will get them eventually. As you ask questions, trust that whatever shows up is useful in some way. Target what you want by asking better-formed, open-ended questions.

If you ask questions such as "Why can't I do this?" you can cultivate and perfect the skill of obtaining completely *useless* data. Believe in yourself as you ask questions. Even seemingly nonsensical imagery means you are getting something *useful*: nonsensical imagery means you are forming a bridge with your subconscious mind. Your conscious mind puts filters on everything you perceive and tries to make sense of it. Therefore, imagery that arises unbidden and may not appear to make sense quite likely represents far more data than is usually available to the conscious mind.

This is what remote viewers in military operations were taught in the 1990s. Whatever imagery they received, they were told to write it down and pay attention to it. The more unique it was, the more potentially important it was. In fact, the methodology taught in Matrix Energetics owes debts to both individually specific dream imagery (Jungian analysis) and classic remote viewing protocol as taught by the Stanford Research Institute and the U.S. Army.

If you begin to trust that you deserve whatever knowledge you require, you can have it. No one is going to stop you. No one will slap your hand down when you reach out in order to know and to be more. Try couching your requests, concerns, and desires in the language of fun; the universe will be quicker to respond to your needs.

I will cover in much greater detail the science of asking life-transforming questions later in this book. For now, trust me when I say that within the question lies the answer.

> *Ask, and it will be given to you; search, and you will find;*
> *knock, and the door will be opened for you. For everyone who asks*
> *receives, and everyone who searches finds, and for everyone who*
> *knocks, the door will be opened.*
> **Matthew 7:7–8**

You can have exactly what you want as soon as you stop trying to control what shows up in your life. Accept the Grace and Love that you are and trust that things are just as they are meant to be. When you accept where you are, you can let go of the emotionally charged state of duality that always wants something better. When you truly let go, the next step in your evolution of consciousness is waiting for you.

Why Can't I Just Make My Problems Go Away?

There are a number of people performing amazing feats of healing in a far more impressive way than I am. They might be domestic engineers or vacuum cleaner salesmen or electrical engineers. Medical training is both a blessing and a curse. As a medical doctor, you are trained to see health and well-being in a prescribed fashion. My first response to the medical method is to ask, "What is wrong with a system that is compelled to share with the patient first what the worst possible outcome could be?"

How would I even begin to answer such a question? Instead, in order to provide a fair and balanced second opinion, why not focus on what could go right for a change? Perhaps some things might get better when we stop focusing exclusively on the problem.

You too can tap into the template of consciousness that is Matrix Energetics. Together, we can create something new and exciting. Every person who reads this book can, and will, become part of that grid of expanding choices and new abilities. How many of you are doctors? Raise your hands. Now put them down. You are not in the classroom anymore. Stop taking notes in the margin of this book. There will not be a test on this material later.

That is not meant to suggest that this material will not test you. Bruce Lee talked about "the way of the intercepting (open) fist." A much more productive approach with this book would be to employ "the way of the intercepting (open) mind." My teacher, Dr. M.T. Morter, president of Parker Chiropractic College, where I went to school, told us to "study everything and believe nothing." It does not matter whether you believe in something or not as long as you use your knowledge and experiences in order to become more of who you really are. *Trust whatever shows up in your awareness even if it doesn't make any sense or appear to be useful.*

Symbolic information is generated by the right brain and represents a potential encoded capacity of around forty billion bits of information. Contrast this with the left brain's paltry calculated sum of seven (plus or minus two) bits of info per second. Some less stingy researchers now put that number a little higher, at forty bits per second.

I want to meet the superhuman grad student who actually counted those forty billion bits per second. Of course, it's not possible to do that; he or she made up that number based on some calculations. But that huge number and all that information make it all sound credible, don't they? The point is that you might want to pay attention to the nonsensical-appearing data, as it may potentially represent far more encoded information.

You do not really know you know something until things start happening around you, within you, and through you—until you experience it. Then you know you know. It is the difference between the Gnostic principal of "I know" and "I think" versus "I wish I knew (I used to know . . . maybe)." *Knowing has nothing to do with thinking, but rather is a quality found in and through the heart.*

I now believe that when we are able to suspend conscious thought, however briefly, the field of our heart takes the reigns. I believe that the heart's wisdom has instant access to the right brain's forty billion bits of information per second. So, based on this theory, *the less you try to think, the more powerful your outcomes will be.*

WALKING YOUR TALK (A BIT TOO CLOSE TO THE EDGE)

Some time ago, I fell off the stage in Vancouver and broke my leg. I was dancing to The Doors song "Touch Me," and I got into a shamanic state with Jim Morrison. (I suspect that he was messed up when he recorded that song!) I forgot there was a stage and that I had a body. I was reminded pretty quickly. The stage was only about three feet high, but *I* felt a lot higher than that at the time.

I fell onto my outstretched leg and heard it pop in a way that a leg should not. I felt a white-hot pain go through me, and my spirit was catapulted out of my body. It was that bad. As well as being horribly painful, it was somewhat embarrassing that it happened in front of three hundred people. On the other hand, if you are going to make an ass of yourself, you should at least make it big enough for Jesus to ride into Palestine on.

Instead of palm fronds, I was waving the white flag: truce or consequences! I had to surrender into a modicum of acceptance concerning my injury or I would not be getting back up again anytime soon. The first place I visited in my mind was where you might expect under similar circumstances: "I want my mommy!" (This, of course, depends upon how good your relationship is with Mommy, but that's where I went.) Then I flashed to a late-night commercial for the Identity Bracelet: "Help! I've fallen and I can't get up!" Where is that darn former surgeon general? There's never one around when you need one! And the next thought was "Hospital!" Those three thoughts went through my head—and they are reasonable points of reference when you have injured your body.

I did not actually know that my leg was fractured until about ten days later. How did I know? The bone pain started to wake me up in the middle of the night. If you are a physician—and I am—there is a test to see if something is broken. You take a tuning fork, give the test instrument a thwack, stick it on the broken bone, and then it hurts like hell! If you are afraid to do a test because you already know how it will feel, then you do not have to do it; some things you just know.

This embarrassing incident at least proved useful. I got to demonstrate for the seminar participants the principles and practices I teach. What better example than becoming a lame horse before their very eyes? I was really hoping someone would shoot me.

I lay there and eventually did exactly what you will be learning in the pages of this book. I "flipped through realities" until I found one where my leg did not feel quite so bad. And then I flipped into a reality where I could stand on stage and continue teaching. It cost me a lot of effort and energy, and I had to limp out of the room with my daughter supporting me that night. The next day I went back onstage, and we performed a group healing. Everyone was worried about me and they actually wanted to help heal me. So this is what we did.

I brought my acoustic guitar out on the stage and we sang John Lennon's song "Give Peace a Chance." But we changed the words. We were singing instead, "All we are saying is give knees a chance." Everybody got into it. We made it funny and flexible, so we did not go to the reality where it was broken. I thanked everyone assembled. After that, my knee started to get better and better.

"DOUBTING THOMAS" NO MORE

I remember sitting in the audience, listening to Dr. Bartlett tell this story of falling off the stage at a previous seminar, healing it, and continuing to teach the seminar—with nothing much to show of the damaged leg only two weeks later other than a small limp—and thinking to myself, "Yeah, right. I just spent HOW MUCH money on this?"

My "healing" experience at that time consisted of a four-week chi-gong class taught by my veterinarian and his wife, and now this seminar that I had flown halfway across the country to attend, taught by a man who sees Superman.

Let's just say that my faith wasn't exactly cemented, and yet despite "logical" evidence to the contrary, I instinctively knew that there had to be something to this Matrix thing. And because of that instinct, doubt or no doubt, not only did I attend levels one and two but also level three a few weeks later in Miami.

By the time I got through level three, my brain was spaghetti. Regardless, I practiced on friends, family, pets, my vet—pretty much anyone who wouldn't laugh at me . . . at least to my face, anyway. But no matter how much I practiced, I didn't have anything solid that I could take at face value as proof that, yes, this Matrix Energetics really did work. A few weeks later, I got my proof in the form of a freak accident with a six-pound Chihuahua named Bitty Bo Bitty.

My afternoon began as any ordinary afternoon might. I was dog-sitting Bitty for my dad when I decided it would be nice to take Bitty and my two dogs for a ride as I drove out to his house in the country to feed his outdoor dogs and cats. I parked, rolled down the windows, and told Bitty and my two dogs to stay in the car and that I would be right back—something I had done a thousand times before.

I looked at Bitty standing in the driver's seat, wagging her little tail and looking at me with love as I got out of the car. I turned, slammed the door shut behind me, and panicked when I heard the most heart-wrenching scream I have ever heard in my entire life. I snapped my head around and saw Bitty's right hind leg, no bigger in diameter than a pencil, slammed in the door! It was bent at an unnatural angle behind her and up in the air in a V-shape, where it was snapped in two and bleeding.

I jerked open the door and snatched Bitty up as I jumped back into the car. I dialed the vet and babbled something about cutting the dog's leg off as I sped out of the driveway. My vet (who was also my chi-gong instructor and who had introduced me to Matrix Energetics) instructed me to "work on her while you drive."

Keeping one hand on Bitty's leg and one hand on the wheel, I began to work on the dog. Immediately, she stopped screaming in pain. I looked at the clock and began to count her backward in time: 4:38 PM, 4:30 PM, 4:15 PM, 4:00 PM ... I asked for anyone who could help me to help: God? Angels? Jesus? I imagined multiple clones of me working on the dog. I tried to tap into the knowledge of the universe by asking questions such as, "What would Dr. Bartlett do? What would Jesus do?"

I tried what I had learned at the Matrix Energetics seminars: windows, modules, and frequencies. I visualized a healing energy ball right at the point that her leg was snapped in two. I two-pointed her hind legs using visualization, since one hand couldn't leave the wheel. With the amount of blood I could feel seeping through my clothes from her bleeding leg and the thirty minutes it took to get back to town, I didn't have any choice but to work on her and hope for a miracle. My nervous breakdown would have to wait.

I visualized her running in the yard that morning and asked, "What if Bitty could run as fast now as she did this morning?" I ran through a list in my head of anything I could think of that would need healing and of releasing pain, adding blood, perfect veins, bones, ligaments, tendons, cells ...

In short, I did everything I could think to do. I arrived at the vet in record time. I'm still not sure if time slowed down or if I was just driving like a bat out of hell. I had on two shirts and a pair of jeans, and all were soaked through with blood when I walked into the vet's office. How much blood could one six-pound wannabe dog have? And did she have any left?

I handed the little dog to the vet, making a point not to look at her leg. The only thing I knew then was that somehow the dog had survived the drive. What I didn't know was if she was literally bleeding to death or about to get her leg amputated and her name changed to "Tripod." I kept my eyes fixed on the vet's face, afraid of what I might see if I looked at Bitty's leg. Now that I was no longer driving and concentrating on Matrix, the panic and guilt set in. I was devastated that I had caused the injury. After *all, I was supposed to be learning to heal, not maim or kill a tiny, helpless dog that trusted me unwaveringly and that I loved dearly.*

"The dog's fine." The vet smiled at me. I screamed, "No, she's not! I just snapped her leg in two with the car door!" He answered, "She's just fine. You did good! We're going to have to get you a job on the rescue squad. You saved her." I still didn't quite get it—even as I watched Bitty now running around the room like nothing had happened, with nothing more to show for her visit to the vet than a tiny spot of dried blood on her hind leg that the vet covered with liquid Band-Aid.

LIQUID BAND-AID? You've gotta be freakin' kidding me! And then it hit me like a ton of bricks: Dr. Bartlett really did fix his own broken leg—just like little Bitty's leg was fixed with nothing more to show for it than a small bloody smear that wasn't even half the size of a dime.

So there was my proof, running around on all fours as if the accident had never happened. As for me, I was covered in blood with my mouth hanging open. Who says the universe never says, "I told you so"?

—VH

The miracle of *my* broken leg was that every time I took a step and it hurt like blazes, I had to *choose to take the next step in a reality where it did not hurt.* Did you hear that? So, first step . . . OUCH! That was a bad step! I really do not want to do that again! "Mommy, can I go back to my room now?" I was saying to myself as I was heading to the stage the next day. The next step just had to be better. How do you make the next step better? By not trying to make the next step better. If you try to make it better, you are tracking and ratifying how bad it is. How useful is that? It's not very useful!

You have a potentially infinite set of possibilities within the morphic grid for what can manifest in the next moment. Imagine with me that in a series of parallel dimensions there are a bunch of different legs laid out before you. One of those pairs of legs contains the reality where you can walk without pain. One way to access the most useful legs for the task is to ask the question, "Where is it?" You have to trust that when you ask, the answer will actually appear.

"Look at," or in some manner "sense," the many possibilities presented. Then feel for the "virtual space" that contains the best solution for you. With a little shift of your *focused intent*, you can choose the best possible outcome and with the least noticeable effort. Focused intent simply means putting something on your mental clipboard and then forgetting about it or letting go of it. If you hold on to the thought of what you want, then you have not released it to the universe. Therefore, how can it be fulfilled? If you ask where something is and nothing happens, you may need to work through a little issue with trust. And, of course, being able to do something like I am suggesting does take a little practice!

How did I know if I had actually accomplished *anything* with all this parallel dimensional posturing? Because my knee would often rotate, it would click, and then my gait would adjust. My next step felt easier and more fluid. Then I would take another bad step. Once you have a reference for creating an imbalance called "broken leg," then you have to keep testing it. The Universe says, "Are you really sure?" In the next seminar a mere two weeks later, I stepped onto the stage and my leg was just fine.

When You Open the Door, Grace Can Walk Through

If a person asks you for help with a problem, what do you do? In my first book, *Matrix Energetics*, I wrote a section titled "The Problem with Problem-Sets," in which I suggested the value in opening your mind and your process to something that perhaps you hadn't even ever considered before. You can learn to open yourself to solutions when you trust—and get out of the way. By choosing to trust first, and act from that trust, you can open wide the door to the field of universal potential. All that is required is that you **pay attention, notice what you notice, and embrace whatever shows up**. By doing this, you allow the principle of Grace to enter into the equation. This can change the odds, even in cases of debilitating conditions.

Revisiting the Problem State

Healing and disease are two parts of the same equation of a closed loop. In a closed system, you can only get out what you put in. I think this is one reason why you only see or hear about miracles occasionally. If you think you are going to learn a healing technique in this book, I am sorry. That is not going to happen. Dissolve the ideas that healing and disease are a polarity and that one can't exist without the other. In an *open system* that is not true. In an open or so-called free energy system, you can get more energy out than you put in.

A miracle is an example of an open system. This subject is germane to everything I teach in Matrix Energetics. When you create and begin to embrace a consciousness model that allows you to transcend the limitations of your physical reality, you can tap into the physics of miracles.

If you have a tumor or a condition, a disease, or whatever, *it has the potential to be healed instantly.* Transformation is a miraculous process that can transcend what you think of as the laws of physics. Here's the point: **miracles operate outside your normal reference for reality.** They happen all the time, but we sometimes miss them because *we habitually focus on what's still the same—on what's familiar and what we know and believe consciously.* We would have a lot more miracles in our lives if we simply began to notice what is different or new!

4

FUNDAMENTAL LESSONS IN DUALITY

You are consciousness. That is what you are. You have chosen to manifest the patterns of information that appear in your personal reality in the form of conditions, structures, families, or finances. At some level, you have chosen everything in your life. Sometimes we actually choose by not choosing. All too often, some of us choose by being in polarity with what we do not want. If we do not want to develop a state of disease, what do we do? We eat right for our health. There is a hidden bomb in that concept, isn't there? If you eat right for your health, what might you be doing? At one level, you are eating to stave off disease.

If you are eating healthy foods in order to prevent disease, you are in an unconscious relationship with what you fear. If you are trying to prevent your cholesterol numbers from going up, you are in a struggle with cholesterol. If you are taking aspirin to prevent a heart attack or a stroke, you are in an unconscious and uncomfortable relationship with all those things. You are in a *paired spin* with what you don't want. In such a configuration, if your fears spin up, your health can spiral down. When did "an apple a day keeps the doctor away" become "an aspirin a day keeps the doctor away"? *Don't buy into all the medical statistics so completely that you become one!*

It is one thing to eat in a way that benefits you because you feel good when you do it. And eating in this manner might coexist with certain principles of eating right for your health. Your personal needs and your energy, coupled with your beliefs, define what being healthy means for you. This is why diet varies with individuals. A person can eat only bacon, lose weight, have a healthy heart, and have his or her insulin levels go down. That diet might represent good biochemistry for that person. Another person could eat only raw salads and do just as well without any other kinds of food. A person from a different planet might consume only raw sunlight and be full of light and energy! However you decide to eat should be congruent with your personal and cultural references for what is healthy for you.

WHY EATING RIGHT CAN SOMETIMES BE WRONG

Health and disease as mandated structures and norms are bad ideas. As I've already suggested, when you are "eating right for your health," you might, unconsciously, be in a relationship, and a polarity, with disease. Bacon could be really great for your arteries. "All right, doctor, if you say so!" No, only if *you* say so! You can say, "I love you, bacon," and it could be just fine to eat it. I am being semi-serious. As a physician, I strongly suspect that much of the nutritional research out there is questionable at best. Much of the so-called health research is actually sponsored and paid for by huge corporate conglomerates, including the pharmaceutical giants.

When the research generates data other than what is desired by these companies, it can sometimes become buried. If a study conflicts with the financial bottom line of the corporate entity, a new study will quickly be undertaken to shift the outcome in favor of the product to be marketed. As the villain, Walter Donovan, in the movie *Indiana Jones and The Last Crusade* said, "Didn't I tell you not to trust anyone, Dr. Jones?"[1] The problem is that if enough people believe a stated conclusion, it then becomes a reality for those ideas that fall into that particular box. That is why I say there is a surprise in each box. What particular box of beliefs have you built around yourself or allowed others to construct for you?

Do not do something just because the experts tell you to do it. Do it because in the moment it feels right and supports your physiology. Do it if it supports your need or belief system as it is showing up in the moment. Do you hear the flexibility in that? Or perhaps as you kneel in prayer at night, you might say a little prayer to Placebo, the medical god of spontaneous healing: "Now I lay me down to sleep. I pray Placebo [a Greek god?] my heart healthy to keep!"

Avoiding the Problem Strengthens Your Relationship to It

If you try to avoid a problem or pretend that it is not there, you run the risk of entering into self-delusional states. This is not helpful. You can acknowledge the problem for what it is, including all the emotional turmoil that may be connected with its manifestation. The charged state that exists when you hold on to the problem helps keep it tethered in physical existence. When you *show respect for the energy of the problem*, you actually diffuse some of the psychic charge off it. Then you can *gently set it aside*, perhaps just outside the threshold to the door of your conscious awareness. This allows you to be neutral about the condition or problem.

Tiptoeing Around in the Medical Hologram

One day a lady came into my office and proclaimed, "You are going to heal me. You are my last hope!" She was obviously in the wrong place! Where did she get the idea that I was God? She said she had tried everything and I was her last chance. I was her miracle.

When I "noticed" what I noticed about her, I saw needles and pills floating in her energy field all about her.

"Forget miracles as a last act of desperation," I advised her. I fashioned an imaginary funnel for her and poured pills into the top of her crown chakra, saying, "Go back to your doctor and trust that the medication he prescribed will help you. Take your medication, because it is apparent to me that you are very congruent with the philosophy and approach of traditional Western medicine. Seek out your miracles wherever they present themselves for you. Trust that whatever shows up in your reality can be useful and may even be tailor-made to your individual needs."

This woman had been seeing traditional doctors for many years and was internally congruent and in agreement with those methods at inner levels of awareness. She needed to recognize this in order to appreciate what had already been done for her. There is a huge morphic field of limitation called "disease"; if you tap into that field, you can, no doubt, create a lot of trouble for yourself. If that is the reality you are playing in, the results can continue to create the disease.

I AM NOT FOR OR AGAINST MEDICATIONS

I am not big on medications and the whole single-bullet theory/shotgun approach to the practice of medicine. I am not against them, either. Because if you are against something, that means you have to be for its opposite. If you are for something, you are in polarity with the thing you are against. I am not kidding. Matrix Energetics isn't about being for or against anything. If you can find yourself somewhere in the middle, these distinctions cease to exist. *Everything, ultimately, is just patterns of light and information.*

I am not suggesting that you deny your cancer or any other medical diagnosis. Do the radiation, chemo, or whatever is the best therapy in the prevailing medical paradigm. Those morphic fields of possibility can heal you if you are congruent with that model at a deep level of belief. Do not pretend to be in a model that is not yours. As long as you play by the rules of materiality (or sin, disease, and death, if you prefer), you define exactly how that reality shows up and you respond accordingly.

Your perceptions about what you think is possible define you. If you need to take a medication or if you require a surgery, just do it. Do not feel like you have failed to achieve a passing grade in Consciousness Raising 101. The idea that you must be free from all adversity still ties you to the concept, power, and morphic field of what you are adverse to.

What you choose to believe is potentially unlimited. Choose to access what you consider to be relevant in the moment. Accept where you are as a good starting point—and then bring in other possibilities. Do not bother fighting against those things that are established. Bring in any possibility that allows it "to be" or "not to be" simultaneously. What is already present, if unwanted, can become something different. Change often occurs when you stop trying to change things.

THE PROBLEM-SET AND BREAKING THE RULES

My goal for you in reading this book is that you realize that all situations in life are merely patterns of light and information. If you want to change anything in your life, change the frequency, density, or quality of the light patterns that make up that reality. Let go of all doubt and *do this with a feeling of certainty.* Try not to second-guess what will be the final outcome. When you begin to do this, you will tap into one of the master alchemical keys for transformation.

Whenever you have a problem, don't always seek out the most obvious solution. Break through your own rules whenever possible. When a rule is broken within yourself, you realize that there was never a rule in the first place. Every rule defines the border and the boundaries of your perceptual bias. There is no reason to play by the rules. Sometimes it is useful to be resourceful by pretending to play by different rules. If your rules are not working so well, borrow somebody else's. You can even make up a new rule in the moment. Do what makes you smile. It is your reality and your life.

Set up a rule set so that when you take your attention off something, that is when the change occurs. When you don't need to do anything, change can occur instantly. Of course, you will do this in such a way that you are still linked unconsciously with what you were paying attention to and trying to change. You are entering into your apprenticeship with the magical "art of not doing." Do not pay close attention. Use *fuzzy attention,* which allows for the power of indeterminacy to overrule any rigidity in the quality and form of your manifestations.

PROBLEM-BASED RESEARCH MODEL

Dwelling on and analyzing a problem can sometimes reinforce its power. If you see problems everywhere you look, then look somewhere else!

Learn to see with new eyes rather than conducting innumerable double-blind studies. Rocket scientist Wernher von Braun said, "Basic research is what I am doing when I don't know what I am doing."[2]

You can research new ways to see and to be. When you surrender to a humble state of not knowing, you can truly be open to learning something new. When you embrace a new perspective, you are literally learning to see things differently.

Behind every problem is a program. Every program creates and is sustained by its own unique morphic field. The morphic field is created by a person's beliefs and expectations and how these match his or her opinion of what "should" be.

Don't wait until you reach a threshold of agony before you are willing to try something different or see things in a new way. People who survive cancer often say it was the greatest gift in their lives. They say that the experience of having a major disease caused them to reevaluate what they thought was most important in life. Why wait for that? Save yourself some trouble, if you can, and begin to reevaluate things right now. If you already have a major medical condition, then perhaps you could learn to minor in some other subject matter. Why make Life's Lessons harder than they have to be?

SUPERPOSITION OF THE PROBLEM
AND SOLUTION STATES

The power of observing from a different reference point is not meant to suggest that you must deny the existence of disease, suffering, or sadness. In quantum potential states, two *electrons* may actually occupy the same space at the same time. This principle in physics is called *superposition*. You can be deeply absorbed within—and completely transcend—your personal Passion Play. In this open state, the *waveform* of your conditions and its inverse (the solution set), or *phase-conjugate pattern*, may cancel each other out, thus preparing the way for a new outcome. As long as you are trying to do something, or even avoiding it, you are not in denial of it; *you are it.*

IT'S TIME TO CHANGE LIMITING PATTERNS OF EXPRESSION

Absorb what is useful in the moment and run with it. When you are able to levitate or walk through walls or wave your hand and watch tumors fall off people, go right ahead and do it. I have heard many amazing stories from people who have incorporated the principles and practices of Matrix Energetics into their lives. I myself have witnessed the disappearance or dissolving of tumors and other "medical miracles."

Does this *always* happen? I wish it were so. If I could heal cancer or anything else with a reasonably reliable certainty, I would do it full-time because I am acutely aware that there is much suffering in this world. I cannot myself heal anything. *Believe it or not, the state of helplessness that we all sometimes feel reveals one of the most important principles in creating quantum miracles.*

DEVELOPING THE ART OF "NOT DOING"

You do not want to enter into the state of needing to do something; that is often when you find you can't do something. Allow the state of consciousness of doing nothing to become available to you and make it your trusted friend. Consciously spend some time and energy constructing this state in advance, knowing that often nothing happens at just the right time and that everything happens when nothing is being consciously done.

The less you do, the more power you can gain access to. When you are trying to do something or make something happen, you are "doing" with your limited awareness of what can be done. When you are in the tide and flow of events, you are no longer resisting their quantum potential benefits. Now anything is possible and many miraculous things become more likely.

GIVE UP IN ORDER TO SUCCEED

In order to achieve success, first give up the idea that you are "the Doer." Instead of constantly doing from the mere human or conscious level, become the "Open Door." Remember to leave your "Light" on so that you may entertain "Angels unawares." When you give up, you are able to step outside of your normal consciousness routines. Doing this allows you access to what you would not normally have, be, or do. The ability to give up and get out of your own way can become your method of choice as the "Court of *First* Resort."

EMBRACE THE CHANGE YOU CAN'T SEE AND WOULD NEVER THINK TO LOOK FOR

If you are looking at a reality where nothing changes, perhaps it is because you are looking through prescription lenses called "nothing changes." If you set up something that you want to see happen, such as a visualized outcome, and then nothing changes, you are in a reality subset called "nothing changes when I look at it." In order to change a habitual experience such as this, shift out of your normal frame of reference. Focus in a new way. Perhaps you could try getting some distance on the problem and then view it as if you were looking at it through a telephoto lens. *Understanding does not predicate or rely upon patterns generated by past ability.*

5

DO NOTHING VERSUS TECHNIQUE

Surveying all the techniques I have studied, I realize that each taught me how to intend for something to happen and then to hang on for dear life to that desired outcome. I was taught that I must do everything just like the Master/Teacher. It must be done this way—and only by these precise steps. If it is not on this piece of paper, it is not real. I can just hear the Wizard of Oz saying to the Scarecrow that he has all the knowledge, but still needs a procedure.

You have to let go for something to happen. The more you let go, the more you let things occur. If you cannot let go, then you will get what you get for the moment and that may be all that will occur. Yes, you can do some really amazingly powerful things in that moment. And the conditions or problems you work on with your plans and your procedures may or may not return. Whatever you do does not matter. Whatever you say or do is just a convenient way to *say or do something to convince yourself that something is happening*.

There is a difference between embracing something so that it becomes you and practicing a technique. When you are trying to do something, you are limited to the reality subset of "what you are trying to do." Now, the subset of what you are trying to do might be very large; I have at least forty techniques and tools on my Batman utility belt. But remember, *your tools are only as real as you are in the moment*.

ALL TECHNIQUES EMPLOY A SPECIAL REFERENCE REALITY OR MORPHIC FIELD

Any technique or method employs a perceptual framework from which its rules of the game are fashioned. This framework creates a unique limited theory of special relativity called "this is the way things are." In other words, somebody had something unique and perhaps even miraculous that occurred in an apparent response to something they "did." Seeking to recapture the magic, they then made up a procedure or a technique. Next, a set of rules was created that were then taught to a group of students.

Enough people then practiced this technique enough times to create a unique morphic field, or consciousness grid. Now everyone who performs that particular technique or method enhances the aura or *morphic field* of that specific system, technique, or belief terrain. This is an example of an *artificially engineered*—or *virtual reality*—template. I will come back to this subject several times in this book.

Matrix Energetics teaches a principle I have labeled the *Quantum Conundrum*. Stated simply, the principle says, **the less you do, the more you have.** Many erroneously think that in order to be a healer, you have to know stuff or be able to do stuff. If we become mired in the idea that we are actually doing anything, *we can get bogged down with our own self-importance.* There are literally billions of complex interactions that are constantly taking place within a human body. It is an absurdity to suppose that we could, exclusively through our conscious thought or intent, really do anything to influence this ultimate masterpiece of creative engineering. It is somewhat sad that we continue to limit our interactions with others to the realm of the merely human. We are the *Door* to the divine, not the *Doer*.

THE LESS YOU DO, THE MORE ACCESS YOU HAVE TO THE ALL

Here is a miraculous testimony from a Matrix Energetics Certified Master Practitioner concerning the art of doing nothing:

I had been suffering from an incurable, insufferable, and previously undiagnosable skin disease for almost seven years. I came to a Matrix Energetics seminar approximately three years ago, hoping that Dr. Bartlett would be able to "heal" me. I had seen more than thirty doctors around the world and had tried innumerable treatments without success or relief of my symptoms. I actually worked in the pharmaceutical industry, so I knew all the top experts in dermatology, infectious disease, psychiatry, etc. No one was able to help me. I could not find a cure for what ailed me. So I came to a demonstration of Matrix Energetics because I had heard about this wacky doctor who was doing some crazy things, and I thought maybe, just maybe, he could cure me of my disease.

I went to see Dr. Bartlett at this free event and he "demonstrated" with me on stage—and I went into a spontaneous backbend (what is known and referred to in Matrix Energetics as Frequency 18). When I came out of this self-generated posture, I felt like I had a terrible headache. Dr. Bartlett asked me, "Is it a headache or just different?" That shift in perspective really resonated with me, and I signed up for a seminar that was scheduled in a few weeks.

I went to the seminar half hopeful for a cure and 90 percent skeptical that anything could possibly change. All weekend long I wanted to tell Dr. Bartlett about my "disease" so he could "heal" me. Instead, a little voice inside my head repeatedly said, "Do not ask Dr. Bartlett for help. The power to heal is within. There is nothing to heal." I listened to that little voice and did nothing. This was a life-altering decision.

Through the course of the weekend seminar, I became congruent with the possibility of potentially having this disease for the rest of my life. I let go. I opened up to the possibility that maybe I did not have a disease after all and that maybe it was just an experience I was choosing to have in a given moment that may have been useful for me.

My disease may very well have been the experience I needed to have in order to move into a new state of awareness, one that is freely available to me in the next moment. I realized, through what Dr. Bartlett was teaching, that it was my choice how I perceived what I was experiencing. I chose to let go of my previously limiting set of perceptions, and in the letting go, I was no longer afflicted with the symptoms. Within a month the disease was completely gone, after years and years and years of unsuccessfully "battling" the ailment.

Interestingly enough, the symptoms, as I had previously defined them, only reappeared in my life when they were useful. What I mean is that they would appear at times when I was not in my own integrity, such as when I was not comfortable in my own skin, with my own power, and with what was showing up in my life. By being able to notice this, I had a built-in feedback system for my own healing and well-being. I began to see that literally through my skin, my unconscious had given me a very powerful gift that enabled me to grow more comfortably into the person I really am. I have not had my condition's symptoms return for several years now.

The point of my story is that people may think they are reading this book for a particular reason (e.g., to get rid of a problem in their life or to get rid of a disease). But there really may not necessarily be a problem or a disease. It may just be something that is showing up in your perceived reality. With this in mind, see if you can shift your perspective on why something might be showing up. Step into it and allow things to unfold. That is where change happens.

—MJ, **Matrix Energetics Certified Master Practitioner**

If you think, then you are going to have to be the one doing the work. That is neither as much fun nor as effective as just letting go and trusting. Let go of the need to think or to consciously manipulate this process. The less you do, the more you can accomplish. *Be guided by the Unseen through the Realm of the Unknown* so that in letting go, you can experience a transformation in all aspects of your reality.

THE LUMINATOR

Years ago, I was surfing the Net and saw something about a device called the "Luminator," which piqued my interest. This is a technology that actually allows you to see distortions in the energy field and then find the "remedy" that will correct the imbalance. Using the Luminator, all a person needs to do is take a picture of the distortion using a Polaroid Instamatic (600 film) and it shows up.

Interestingly enough, using a digital camera does not work because digital auto-corrects. This is the same principle in action when you are experiencing something "unusual" and your conscious mind says that nothing is happening and auto-corrects to the state of nothing happening. It brings you back to your conscious focus, which says, "I am a body." You are not a body. You are a body of information.

Eighteen years ago, an engineer and scientist named Patrick Richards created Bio-Liminal Photography and invented the Luminator machine, which consists of a plastic tower that has six glass rings filled with water in it set at different phase angles. It has a big fan on the top and a big fan on the bottom. That's all it is. He was attempting to create a device that would address the "sick building syndrome": office workers manifesting vague and unspecified illnesses in buildings that were essentially closed air circulation systems, receiving no fresh-air ventilation. His idea was to improve the quality of the recycled air by eliminating the thermal layers in the room and enabling the efficient distribution of heat.

The Luminator was designed to balance air temperatures from floor to ceiling and from wall to wall for efficient energy management. Richards found that after its installation in an office, staff members began reporting that their general health had improved, including the reduction of lower back pain, eyestrain, stress, and migraine headaches. On further investigation, Richards discovered that as well as balancing the room temperature, the Luminator altered the magnetic field and changed the available light in the room from incoherent (going in all directions) to coherent, or polarized, light. By removing the thermal layers, you ionize the room and the photons become super-coherent.

When you have a super-coherent external environment and you bring something not so coherent into it and take a picture, the mismatch shows up as distortions on the film. Richards discovered when photographing individuals within this field that the photons were either clear and crisp (coherent) or fuzzy or fractured (incoherent). After years of further research, he concluded that the photographs were revealing the quantity and quality of the cell light being emitted from individual subjects. A person with strong vitality and lack of inner stress, for example, emitted more photons, thereby creating a more coherent image than an individual with less vitality and greater inner stress.

Once I acquired the Luminator technology, I really began to walk my talk, and I really began to redefine what Matrix Energetics is about. Using the Luminator, I could get instantaneous "real-time" feedback about the effects of what I did or didn't do in a particular clinical situation. One of the most significant revelations was that the less I did and the less I concentrated on achieving a particular outcome, the more beneficial and powerful the changes in the coherence of the person—as indicated by the changes in the photographs. After seeing this feedback, I began to relax and really live what I had been teaching in Matrix Energetics. I began to embody the art of doing nothing.

Below are some photographs of clients taken with the Luminator that clearly show the mismatch of coherent and incoherent energies. My process for taking the photos was consistent: I put an X mark on the floor in my office right against the wall where the patient would stand. I then moved back about eighteen feet and marked another X on the floor. That is where I put my stool, always in exactly the same place. I would then take the picture.

EXAMPLE 1: BEFORE (BASELINE)

This woman felt depleted and sore all over. She is a massage instructor, and she had been to seven practitioners before she saw me. Notice that even the background is distorted in the picture, although I did not move the camera. (I do not use a tripod, as it is too much of a hassle with changing the film and other parts.) The scenery is distorted because the field around her is distorting what appears to be the room itself.

The first thing we did after taking the first photos—the baseline photograph—was to make an imprinted remedy. An imprinted remedy is something we can thank German innovation for. The Germans determined that homeopathy was being corrupted by all the electromagnetic frequencies, such cellular phones, telephone lines, and modern technology. Homeopathy is just an electromagnetic signature, and there is nothing physically actually remaining in the molecules. The Germans understood that the more diluted the formula (the less you do), the more powerful the outcome.

What the Germans did, once they realized there was nothing remaining in the homeopathic remedy, was use lasers and holography to actually imprint the remedies onto software. We are not talking about imprinting radionically; we are talking about the actual imprint of information. Once they imprinted this, then they would place the remedy on a magnetic strip, and the patient would wear the remedy.

So with this patient, I found a "crown chakra remedy" and applied it to her accordingly. When I saw the next Luminator photograph, it was apparent that the crown chakra remedy appeared to do a pretty good job, so we added a universal healing frequency (taught in the seminars) into the crown chakra, and we got this second picture below.

EXAMPLE 1: AFTER MATRIX SESSION FREQ. 3 INTO CROWN CHAKRA

The change is pretty definite. Can you kind of sense the energy radiating out? She felt tremendous. All her symptoms vanished. It took less than a minute.

Here are some additional examples of the Luminator photos that demonstrate coherent and incoherent energies as well as the value of doing as little as possible.

EXAMPLE 2: BEFORE (BASELINE)

Female Client: Obvious phasing distortion on Luminator photograph.

EXAMPLE 2: AFTER MATRIX SESSION

Some distortions still present but much improved.

EXAMPLE 3: BEFORE (BASELINE)

Client's husband: Obvious distortions present on film.

EXAMPLE 3: AFTER

Here is the follow-up Luminator photograph after the wife's treatment. It is important to note that the husband was not treated at all. These photos demonstrate the very real effects of quantum entanglement and the phenomenon of energetic rapport.

EXAMPLE 4: BEFORE (BASELINE)

Young male client experiencing a headache when this picture was being taken.

EXAMPLE 4: AFTER MATRIX SESSION

Post-Matrix session (five minutes later).

EXAMPLE 5: BEFORE (BASELINE)

Female client primarily complaining of feeling really out of it that particular day.

EXAMPLE 5: AFTER MATRIX SESSION

Client reported "feeling like myself again!"

LESS IS MORE

I learned from taking before and after shots with the Luminator that less is more. Whenever I would attempt to fix the distortion I saw in the photographs, my "after" shot invariably contained more distortion. Paradoxically, when I did as little as possible and let go of the need for a particular outcome, the distortions in the after shots were generally less distorted and the person's symptoms would improve greatly or vanish as well. This absolutely revolutionized the way I practice Matrix Energetics. Remember, *the sense of struggle creates the struggle.* Let go and trust that "it is done," and it usually will be.

6

THE QUANTUM CONUNDRUM

When I first started working on this book, I thought about titling it *The Quantum Conundrum*. The more material on alternative science I read, the fishier some of the basic scientific assumptions of quantum physics and its kissing cousin, the *general theory of relativity*, appeared. I am beginning to suspect that neither theory is correct. Based on objective physical science data from a number of sources, it seems to me that *we have manipulated the physical data to agree with our cherished theories*. There are some very disturbing scientific studies that suggest that quantum theory and real life disagree by a factor of ten to the fortieth magnitude. That is a pretty big "missed" understanding!

The quantum conundrum, as I stated previously, is that *the less you do, the more you have*. This is opposite to how most people live their lives. Most of us have been taught that if we want to get ahead, we have to work harder and longer. I worked as a doctor for twenty years, often six days a week. I can tell you without hesitation that the more you work, the less you have.

To work smarter, not harder, isn't much better. It makes no difference how smart you are; it's really about how creative and innovative you are with your time and energy. So if I were to put this in an equation, it might look something like this: energy multiplied by creativity, divided by time, equals Innovative output: $ec/t = I$. This means that *when you find ways to use more energy in less time, innovative dividends become the well-deserved result.*

> *If A is success in life, then A equals x plus y plus z.*
> *Work is x; y is play; and z is keeping your mouth shut.*
> **Albert Einstein**

MY RULES: THERE ARE NO RULES, ONLY SUGGESTIONS

Things cannot change if you are holding a state of "no change." If you are holding and observing only what is wrong, your concentrated focus can reinforce the problem state or condition. That brings me to my set of rules, which are really more like suggestions. They are not really rules.

The first thing you do is *drop down*. What does this mean? We are used to being in our heads; this is where we live most of the time. We think that our thoughts control our reality, or at least our perceptions. Actually, thoughts have nothing to do with reality whatsoever. Our thoughts keep us bound in a prism and a prison of what we believe is true. Our perceptional reference frame molds and shapes what shows up for us in the moment. When we stop asking questions and when the mind is silent, we are no longer in our head and we can *drop down* into our heart and into a larger reality.

Imagine a pebble that is tossed into a pond. The waves ripple outward in ever-expanding circles. Now imagine a pebble dropping down your throat and into your chest. You may experience the waves unfolding out from your chest, like the waves in a pond, and into your energy field. From this larger-reality point of reference, there are many more possibilities and realities to perceive.

GENERATING THE ALL FROM NOTHING

One of the greatest secrets of so-called spiritual alchemy is the ability to place intent—and then let go and do nothing.

Intent can be defined as the creative act of using the many and varied parts of your total conscious experience to define a set of new experiences, realities, or outcomes in your current experience. In order to do this effectively, you must focus your imagination to create a new sensation that will initiate a flow of subtle energy to directly or indirectly influence or manifest the desired events and effects. Thus to create, focus with feeling.[1]

The key ingredient, however, is as stated: after placing intent, let go and do nothing. If you do this, you create a void that then can be filled with your heart's desire. It has been said that "nature abhors a vacuum." One of the ways to fill the vacuum is to look for and seek out examples of others who have mastered what we desire to accomplish or to understand. These individuals are always present in various aspects of our lives as a conceptual mirror for us. You can ask, "Who seems to embody the principle or thing I wish to understand?"

HOW TO SHIFT YOUR REALITY REFERENCE FRAME

When I am doing Matrix Energetics with someone, I shift the focus of my perspective. I have trained myself to allow the focus of my attention to be gently and unconsciously shifted and redirected. I am continually scanning my external and internal environments, looking for subtle clues. This softly held scanning focus allows me to notice whatever catches my attention in the moment. Imagine that your eyes are constantly taking in information, like a computer. As you do this, do not attempt to analyze or judge the information presented to you.

I am allowing the boundary state between what is "real" and "imagined" to become a little blurred or less rigidly defined. Within this blur there are motion, possibility, and unconscious form. In the act of consciously observing "the blur," my conscious mind/left brain slows it down in order to perceive what I am looking at.

A rule I have adopted is that I don't have to consciously see what I am working with in order to interact with and change it. The activity of conscious observation breaks down what is essentially a motion picture into innumerable individual slides.

The conscious mind can only pay attention to one observational frame at a time. This neurological process represents the "doing state" within the problem-set, as in, "There is something wrong, and I have to *do* something about it." This literally determines the difference between the act of observing and the process of interacting.

The word *process* implies that there are rules, guidelines, and machine-like behaviors involved with what you are doing. A process, once committed to, is a set of least-action pathways. A *least-action pathway* is a neurological response to learning and then embedding a behavior into the unconscious realm of a habitual response.

Some examples of least-action pathways could be things like learning to brake your car when something rolls out in front of you or returning a serve when playing tennis. These responses, when committed to unconscious reflex, allow us to deal instantly and appropriately with threats or emergencies in our environment. Once we commit these activities to a program, our behavior in similar circumstances becomes immediate and somewhat machine-like. This is a good thing that allows us to respond unconsciously and fluidly, like a well-oiled machine, when appropriate.

For instance, if you have ever studied any martial arts form, you practice a particular punch or kick thousands of times so that your unconscious mind eventually takes over and directs the response. It is very important to not have to remember how to block or punch in a street fight! Of course, a better and more useful least-action pathway is to notice the first signs of danger in your environment so well that you never put yourself in harm's way in the first place.

Least-action pathways are useful for many activities, but not for Matrix Energetics. Every time I look at something consciously and attempt to name it and change it, I am applying the same old tired and worn-out model of Newtonian physics in its emphasis of forcing physical reality to conform to the constructs and expectations of a personal or world model. In other words, the way we notice and choose to construct our reality often constitutes a multitude of least-action pathways in our neurology.

Building the Framework for a New Reference

When you apply a different *reference frame*, different information filters through into your conscious awareness. Our reference frame, or view of reality, is just something we are making up. An analogy would be taking lenses and changing or rotating them, like in the movie *National Treasure*. In that film, Ben Franklin had invented a pair of spectacles with multiple, interchangeable, colored lenses. Each lens set allowed the viewer to unlock different information otherwise invisible to the naked eye. The green lenses, for example, would always present certain specific types of information or patterns. The red lenses revealed additional data to your senses that previously had been suppressed by your conscious mind.

Whichever lens or viewpoint you choose to utilize will largely determine what you are able or even likely to see. If you want to see more or see differently, apply the same principle to your conscious mind as it prioritizes and filters "useful" information. You can decide to implement a new rule that you will notice, say, only 1 percent of what you would have previously considered irrelevant and ignored. You can choose to be more aware of what was, previously to you, largely unconscious. You can adjust the rheostat of your mind to let in more light and information.

Every model of reality is based on our best-guess perspective. There is a decided difference between the rules for how my reality works and how your reality works. How do I know that? You are you and I am me, and unless our experiences collide in a seminar together or in the pages of this book, we are separate. That does not suggest that either reality frame is in some way better. It is conceivable, and even likely, that in key instances my approach would be a more useful reference in a particular set of circumstances than yours and vice versa.

If you build houses for a living, for example, your practical experience as a base of reference in that field is likely far more appropriate than mine. If we were both to attempt the construction of a tangible source of shelter, odds are your version wouldn't suck! If I were to build a house, it would likely be full of odd angles and potentially a very leaky edifice. If we are framing a house, constructing a unique sentence structure, or even laying the foundation for the structural components of a new reality, the skill sets are different. In each of these examples, the creation of sensory-specific and individualized reference frames is relevant to the beliefs and contextual reference of the beholder. Not only is beauty in the eye of the beholder, but everything else is as well.

THE DOORS OF PERCEPTION: BREAK ON THROUGH TO THE OTHER SIDE

I believe in a long, prolonged, derangement of the senses
in order to obtain the unknown.
Jim Morrison

You alone can ratify the meaning and quality of your experiences and give them your stamp of authenticity. Because we are different, there is a difference between my perception of the way things work and yours. We can choose, through mutual points of agreement, to *construct a special relativity model* that encompasses or joins together what you and I choose to believe or make up.

We can choose to be in rapport and create a mutually shared reality where we decide together what the rules are. We may then formulate what classes of perceptions will drive those rules. Of course, we are free to formulate unique ways to determine the meaning of our experiences within our specially created reality subset. Matrix Energetics is an example of just such a special-case reality subset structure, because I made it up. Together we reinforce it by choosing to share in a set of mutually held beliefs and experiences. We have created and given life to Geppetto's quantum puppet Pinocchio.

With enough individuals sustaining the consciousness of our special case reality, such as Matrix Energetics, it functions consistently and reliably. When this field of consciousness reaches critical mass, a unique morphic field resonance has been created. *Once this threshold is achieved, the system of thought or belief becomes self-generating and sustaining: "It's alive!"* Love your dreams enough that you begin to embody them. I have!

In order to be able to change a long-standing condition, pattern, or behavior, it is important to *learn to focus your awareness on the information or state that has a feeling response attached to it*. When I talk about feelings, I am not equating them with emotions. I am referring to becoming aware of a feeling response, which starts simply by *noticing what you notice*.

DOORWAYS OF PERCEPTION EXERCISE

Let me give you a metaphorical example. Imagine that you have an unwanted pattern of behavior you would like to change. Now imagine that before you are ten doors lined up in a row. Each one of those doors, when opened, contains some pattern or type of information. The information behind an individual door may or may not be relevant to your needs. The way to discern if you want to go through a particular door will be based on what information you get—what you notice—but not how you feel.

When you open the first door, you feel or sense nothing, so you close it. The second door you open has a warm feeling exuding from within and there is a bright but soft light shining there. This is a definite possibility, you think—mark that one for possible further exploration. When you try the handle of the third door, it won't open for you. Unconcerned, you move on to the next door.

The fourth door is a soft green color. The frame feels rich and smooth; you notice you are more relaxed than you were a moment ago. Just touching this door generates a feeling of warmth and welcome deep inside you. You get it. This door has your name on it, and you decide. Without hesitation you somehow know that you could walk through this door and find whatever answer or help you need.

Somehow the green door resonates unconsciously with the terrain of your deep inner nature. That's what I mean by "feeling it," not by being in an emotional state. If you were to analyze or calibrate the reasons for your choice, what are some clues you would notice? Let's start with your physiological responses. "Hmmm . . . I feel calmer inside, like my anxiety level has dropped way down. My heart rate is slower and steadier. I seem to be breathing more slowly and more deeply. I feel very relaxed yet somehow inexplicably more alert and refreshed."

All this physiological change resulted from just thinking about moving through the green door. The physiological clues provided are definitely helping you make a good choice in the moment.

We learn to make sensory-based generalizations about the nature of doors. We form an unconscious database about what we discover each time we encounter a new one. Over the passage of time, having seen and gone through many doors, we create a *least-action pathway* where doors are concerned. That way, even if we encounter a submarine hatch or a hobbit hole, based on our previous encounters with doors, we can still figure out how to move through it. This is necessary so we don't have to rediscover the principle of doors afresh every time we encounter a new one! In Matrix Energetics, when you *notice what you notice* and respond to the clues provided without questioning or thinking about it, you have made a good choice in the moment.

The other way to deal with new information is to open your eyes and mind and accept whatever you see, feel, and experience in the moment. **Trust that whatever shows up is useful in some way.** In order to create a new reality or to bring miracles into our lives, we can begin by learning to see the things in our lives in a new way.

We create the aspects of our reality that we come to accept as a part of the fabric of our beliefs, expectations, and experiences. We, to no small degree, make up the tapestry of energetic patterns that we call "our world."

7

THE SCIENCE OF MADE-UP STUFF

In scientific research, each new approach to reality generates more mathematics. When approaches are tested and experimented, the physicists and math geniuses find things that do not match their model. No problem; they decide to just generate more equations and theories. When experiments are inconsistent with the data, they either fudge the data or create new equations to match their experimental results.

For example, if you take a piece of matter and accelerate it to the speed of light, according to the equation it becomes infinitely heavy the faster it gets. We don't know if this really happens, but the equation $E = mc^2$ suggests that it should. So if the results of the equation don't balance out on both sides of the equal sign, or the real-life results don't make sense, scientists just take one part and cancel out the other part so that it comes out OK. I'm not making this up. Mathematicians even have a name for this process: *renormalization*. Quantum physicist Richard Feynman called it "dippy hocus-pocus."[1]

THE UNIVERSE CAN REORGANIZE JUST TO BE CONGRUENT WITH OUR IDEAS

If you create a reality that says "this plus this plus this equals no miracles in my life," then that is the equation you work from. If you work from that equation, it is no more real than $E = mc^2$, which, by the way, was not even in the equation Einstein submitted in the first place. Because Einstein was dyslexic, he had some terms backward.[2] Einstein was so dyslexic that he stuttered. As a little boy, he stuttered so badly that he developed a habit of saying a sentence first in his head and then saying it out loud. Sometimes, though, he forgot whether he had said it on the inside before he said it on the outside, so he would often say it out loud twice. Because of this quirk, people often thought he was dense or stupid.

For many people, their concept of Einstein is akin to an archetypal image of God. Einstein had the white hair and the whole "bigger than thou" thing going—even the unzipped trousers. He simply wasn't paying much attention to this world. He was somewhat like John von Neumann, another famous mathematician, who would claim that trees stepped out and hit his car as he was driving. (He liked to drink.)[3] This is all true. You can do the research if you like. But I have done it for you. You can trust me on some of it. Or maybe even all of it.

If you set up a rule or if you set up a way of observing something and then you set up the rules for how it is observed, then you are always within your rule set. So when Einstein said that nothing could travel faster than the speed of light, what he meant was that nothing could be observed to be traveling faster than the speed of light because the tool we make our observation with is light itself. Therefore, light cannot travel faster than itself. This does not mean that thought does not travel faster than light. This does not mean that *torsion fields* Soviet physicists worked with in the 1980s are not much faster than light. The problem is that once you analyze something and break it down, then you must observe it from your rules for reality.

There are two processes our brains use in order to perceive, transform, or classify information: *serial processing* and *parallel processing*. Serial processing is linking one thing to another to another to another. It is like differential equations in trigonometry class—not that I would know, because I flunked algebra. The only reason I became a doctor of chiropractic is because the program didn't require me to take any physics classes that required complicated math. Parallel processing involves at least two processes occurring simultaneously, and within the body, it is the neural integration that underlies complex mental processes.

> *Do not worry about your difficulties in Mathematics.*
> *I can assure you mine are still greater.*
> **Albert Einstein**

Even the great Einstein occasionally had problems with math. He would come up with these equations that made no damn sense. But he would get the right answer. Or he would have the right equation and the wrong answer. He was dyslexic, remember. So what Einstein would do is piece together the best parts. He was such an important figure with the crazy hair and all, an icon, that people just tended to want to believe him.

One of the postulates of special relativity is that nothing in our observed universe can travel faster than the speed of light.[4] Well, do you know why you cannot go past the speed of light? Because light is the tool you are measuring with! This is the reason. Part of the electromagnetic spectrum includes light. If light is the tool you are using to perform your measurements, then all your measurements of velocity are performed in relation to that factor. That is why Einstein said nothing travels faster than the speed of light from a relative point of measurement. He never meant to suggest that you cannot go faster than the speed of light. You simply cannot *measure* it.

SPECIAL RELATIVITY OF CONSCIOUSNESS

Whenever you artificially describe something with enough detail and conviction, you create a warp bubble or "special case relativity" where it exists. The science is "real"—until it's not real. The model of your reality holds true until your experiences contradict it too much. When met with experiences that are contradictory to your model, you will either suppress the information into your subconscious or form a new model that includes the contradictory experiences.

In order for us to interact, we each have our relativity of consciousness warp bubble. You have yours, and I have mine—and then we can create something unique if we agree together. In order to interface with someone else's reality bubble, what you can do is allow for the creation of a unique bubble of reality that encodes for your experience and theirs as well, and creates a new special relativity where both rules apply or don't.

In other words, things that are useful can stay and things that are not can cancel each other out. That allows us something called *energetic rapport*, where special kinds of things can then occur. When we are connected in this way, we have created a special unified field of consciousness. Actually, what we have done is create a special relativity that through the heart accesses the unified field of consciousness. At the subatomic level, physical reality itself is reduced to a field of probabilities that can be manipulated by the mere act of observation. Minute changes in momentum and trajectory of subatomic particles can propagate to cause far greater physical effects in the phenomenal universe.[5]

PARTICLE PHYSICS AND THE
INDOCTRINATION OF SEPARATION

The building blocks of our concrete world belong to the realm of *particle physics*. Do you ever wonder why it is that scientists keep smashing the atom into smaller bits? The number of new particles that are created and discovered from this process of scientific deconstruction is amazing. We break things down, look at the scattered pieces, and assign a name and function to them. Have you ever taken a toy, such as a little red fire engine truck, and then smashed it with a big hammer?

If you hit a thing hard enough, parts scatter everywhere. If the fire truck was plastic, it shattered into little fragments. What if you have all these little fragments and you name them all: that is a prusson, this is a lickamajick, this is a Klingon, and so on. Have you ever tried to put these pieces back together after you smashed them? It is pretty difficult to do, if possible at all. Of course, your mom or dad probably just called your experiments a mess!

I think things are simpler than that. I think that everything is just spirit and whatever you name it becomes that. So yes, maybe *virtual par-*

ticles did not exist until they did. Once we thought of them, however, then it's too late to take them back. A scene from *Ghostbusters* makes this point. When they were trying *not* to think of anything at all, what showed up was the big Stay Puft Marshmallow Man, which actor Dan Akroyd's character defended by saying, "I tried to think of the most harmless thing. Something I loved from my childhood. Something that could never ever possibly destroy us. Mr. Stay Puft!"[6] Well, your beliefs are like that too, except maybe not quite so sticky.

Why do we think that tearing something to pieces and then analyzing its parts will tell us how it really functions? I remember dissecting a starfish in high school biology lab. I didn't bother with the instruction sheet; I just took it apart with my knife. I learned nothing from this experience. Fortunately, I did not progress to dissecting the goldfish I had at home. I had a far more exciting encounter with a starfish at SeaWorld when I gently cradled one in my palms. I could feel and appreciate the unique beauty of the life-form nestled within my hands. I cheered along with everyone else in the theater when Elliot in Steven Spielberg's classic movie *E.T.: The Extra-Terrestrial* saved the frogs in biology class from imminent dissection. This led to the formation of "physicist" Spielberg's theorem, "Save the Frog—Kiss the Girl!"

The error that occurs within a strictly mechanistic framework is when we see and classify. We seem to think that by breaking an organism or an idea down to its component parts, we have understood what it is really about. There is a huge difference between smelling the odor of formaldehyde while stooping over the dissection tray to peer at a lifeless amphibian and catching and holding a live frog. Have you ever caught several frogs in an enclosure and cheered them on as they "raced" for the finish line? I did this with terrapins in my backyard when I was a young boy. It's not as exciting as watching frogs leap for the goal's end, but you get the idea: you learn more about what constitutes life by actually observing living examples.

In Matrix Energetics, we teach people how to turn frogs into princes. It is the difference between accepting someone's model and actually experiencing it for yourself and realizing that it is not at all what you have been told. It is the difference between stagnant theory and vitalism.

Scientists deal with the classification of inert nouns. "Phylum and forget 'em," I say. *Life is a process.* The eminent physicist Werner Heisenberg said, "Atoms are not things."[7] Physical science writer and physicist Nick Herbert extrapolated from Heisenberg's original statement when he said, "People are not things in the same way that atoms are not things."[8] A body or person is composed structurally of atoms, the conceptual building blocks of all nature. The atom is composed of electrons and the nucleus, which consists of protons and neutrons. But get this: an electron doesn't actually exist except for when it is observed. When we look at it, only then does the electron assume the semblance of a fixed and stable orbit.

When not being observed, *the electron exists in the form of a probability cloud.* This cloud is composed of all the probable orbits it could be in before the act of measurement fixes it into a stable orbit around the nucleus. Since we humans at the atomic level are composed of atoms, *we can be said to consist of a series of states composed from an infinite set of oscillating possibilities.* The ability to let go of our fixed attitudes—our reference frame—can elevate our perceptual platform from which we judge what is possible and what is not allowed. A little flexibility of thought can do much to alleviate or at least mitigate much of what we find in the human condition.

It is important to have fun, because when you are having fun, you start to make virtual particles called "fun stuff" that are really useful. Virtually and literally, *you start to have a good time.* Why is that? Because when you make virtual particles called fun stuff, you actually start to change the orbital spin of the electrons that make up the patterns of neurochemistry in your brain and nervous system. When you start to make better neurotransmitters, you can more readily experience altered states of fun and joy. And you might experience *endolphins* as well as *endorphins*!

Do you know what theoretical physicist John Archibald Wheeler (Feynman's teacher) said about the universe? He called it "a meaning software," located "who knows where."[9] Now, how many of you think he was both confused and not confused? He realized that the inherent order he saw in the universe went way beyond any probability that it was put there by chance. Allegedly, a number of years ago, Wheeler had a health scare that included a near-death experience. Wheeler left this body and "experienced" the universe. When he came back, he said there was something he had to share because he might die. His takeaway message was "if there's one thing in physics I feel more responsible for than any other, it's this perception of how everything fits together."[10] I think that he apparently realized that the mathematical models do not get to the heart of the matter, which is the awareness of the many as the All.

> *As far as the laws of mathematics refer to reality, they are not certain; and as far as they are certain, they do not refer to reality.*
> **Albert Einstein**

8

THE POWER OF COMMAND VERSUS CONTROL

Whereas *Newtonian physics* emphasizes force, *quantum physics* emphasizes finesse. There is a big difference between control and command. Control assumes that you know what is going on, that you have all the factors, and that you can do something against something in order to make something happen. This is that analytical left brain saying, "I can do this." No, it can't. Excuse me—it can't.

Command issues forth from the heart. To command Life is your birthright. When you observe your life in this way, without judgment, *collapse of the wave function* occurs in ways that are not based upon the probability of what usually happens. They instead reconfigure around the thought "What could happen in the next moment?" Does that sound useful? This is physics too. It really is.

Command does not involve control. When you are in command, you are not trying to make something happen. Have you ever read the passage in Isaiah in the Bible where it says, "Concerning the work of my hands command ye me"? The concept of command has to do with stepping into a space of the heart where you are unified with the field of all possibility. From that hallowed and sacred space, when you say, "Let go," *it is done*. Words become virtual cups of light, patterns of information. The way that they are woven literally creates a magical *node of possibility* that opens up the *dimensional capabilities* of your awareness to experience other states, other things, and other places. This concept will become increasingly obvious to you as we continue through this book together.

> *Thus saith the LORD, the Holy One of Israel, and his Maker,*
> *Ask me of things to come concerning my sons, and concerning*
> *the work of my hands command ye me.*
> **Isaiah 45:11**

DROP DOWN, PLACE INTENT, AND LET GO

I want to emphasize something very important. If you were to come to a seminar, you might observe things that frankly look a little like a quantum evangelical revival. What you would apparently see me doing doesn't really involve the state of me doing anything at all. This is important because *the act of doing creates resistance to what could happen in the very next moment.* In order to do something, I have to gear up to do *work.* If I am going to dig a hole with a shovel, what do I have to do? First, I have to have a shovel. Then I have to decide where to dig. Then I have to feel the ground and feel how hard the ground is. Then I have to calculate how much *force* I am going to need to use in order to make my hole. Does this make sense? All this is conscious calculation.

It is much different when you just *drop down* into the experience. Then there is nothing to do, nothing to be. *There is nothing to change.* It is that state of merely noticing what you notice without providing judgment. That free-flowing state is so different from "holistic health," or disease, or disease care, or technique, or conditions of treatment, or conditional analysis of conditions, or even transactional analysis of conditions (trance-actual, for you psychiatrists out there reading this who have already diagnosed me).

There is no difference between states of matter and matter-of-fact states. It is a matter of the *informational content of the field* or spin of the electrons that causes something to react or respond in a particular way. The act of observing something instantly changes it at the quantum level. You are composed of light patterns held in informational templates; this is what you are. Learning Matrix Energetics is not only easy, but it represents concepts and abilities that are fundamental to your reality. It is your attempts to do something, your need to observe something, and your desire to be powerful that keep you from experiencing those things. *Drop down. Place intent. Let go.*

Drop down implies entering a state of altered possibility where other rules, or no rules, can then apply.

Place intent is a concept that is very often misunderstood. We think intent means that we decide what we want, and then we try really hard to get it by doing affirmations or visualizations. Does this sound familiar to you? Doesn't it also sound like a lot of work? And why do you have to do it so much? Because you do not think you deserve it. Because you do not think you already have it. You are already in the relationship with the things you want. You do not understand that they are indivisible and cannot be separated. It is your desire to have the thing that amplifies the state of not having it. Does that make sense? Play with intent as you would play any other game; intent is simply another virtual reality simulation.

Let go is really quite a straightforward concept: you have to let go for something to happen. The more you let go, the more you let things occur.

Bruce Lee taught three key principles or rules to his students: (1) *Absorb what is useful.* That means to notice what you notice in the moment. (2) *Discard the classical mess.* This means what I was just talking about— about smashing up all these particles and then trying to put the universe back together and make sense of it. It does not make sense. Discard the classical mess, the rules that are proscribed in your reality, and your pre-scription for the way things are. Finally, he said to employ the third rule: (3) *No way as way.* This means that in the moment, you want to apply the rules that show up, but in no way limit your self-expression to only those rules. Entertain other patterns or possibilities as well. You can become a shape-shifter like a shamanic chameleon. You can learn how to transform your reality by breathing in one reality set and breathing out another. It is not as difficult as it sounds; and if it does not sound difficult, that is because it is not.

EXPAND YOUR QUANTUM CONSCIOUSNESS

One of the useful ideas in quantum physics is that this is a participatory reality. We are making stuff up. The way we choose the perceptions or the parameters of our reality defines what we are able to experience in the moment. Now, here is the problem. If things have to make sense, then you have a very narrow, limited set of possible outcomes defined for you. What if things could not make sense and make sense at the same time? When you start opening up the parameters of what is possible, you allow for your normal expectations as well as that little bit extra.

QUANTUM PHYSICS AND COMPLEX NUMBERS

In physics, in order to describe the collapse of the wave function, scientists use what are called *complex conjugate numbers*. This complex conjugate, when multiplied by itself, always produces a real number as the product of the operation. For example, take the complex conjugate $(3+i)(4-i) = 12$. The imaginary number, i, is understood to represent the state of all the possibilities that exist beyond the domain of time and space before an act of observation occurs. In the act of observing and from a state of all possible outcomes, one specific action occurs. This result, in our con-sciousness domain, represents what we notice or what appears to be the end result.

ALLOW FOR GRACE IN YOUR LIFE'S EQUATION

Those of you who have read my first book have heard the story I'm about to tell, and I want to add some quantum insights I've had around this strange event that happened to me.

∴∵

It was just after four o'clock in the morning on a bitterly cold January day in Bozeman, Montana. The wind was blowing hard and a solid sheet of glistening snow was steadily falling. It was not an ideal day for a six-hour road trip to beautiful, rustic Missoula. But my contacts there had booked a solid weekend of clients. I was quite likely to make more money during this weekend than I had made the entire previous week—and we really needed the income. In my previous career as a professional musician, I never missed a single gig—and I wasn't going to start now. The show must go on. I wearily pulled on my jeans and sweater and went to the closet to get my heavy coat and snow boots.

As I was leaving, my wife called out to me, "Be careful of black ice!" I had never seen black ice and didn't believe in it. Shrugging off her caution and resolving not to be late, I put my foot down on the gas pedal and my vintage 389 c.i. engine and surged uncertainly down the deserted highway. I was glad that the roads were empty and anticipated being able to make up some time on the long stretches ahead.

I barreled headlong toward my destiny at a speed in excess of 80 miles an hour. With the great time I was making, I'd be in Missoula soon, I thought. Just outside Butte, I encountered the very phenomenon that my wife was always worrying about: the legendary, slippery, and all-but-invisible black ice. Not only did I belatedly discover the reality of its existence, but the patch with my name on it was on an icy bridge just outside Butte's city limits. My tires started across the patch of slippery death that had formed in the center of the bridge. Horrified, I felt my wheels begin to skid out of control. In a panic, I took my foot off the gas and gently pumped the brake pedal, but I was moving too fast.

That old car of mine just had one of those front lap straps as a seat belt. There were no fancy air bags or antilock brakes and virtually no chance I was going to survive this oncoming crash, barring divine intervention. I frantically applied my brakes more forcibly, fishtailing the back end of my car, so that I was now racing head-on for the bridge pylons. I looked down at my speedometer seconds before impact and noted that it showed a crisp and lethal 65 miles an hour. I was staring death in the face and it was grinning back at me. Accepting my fate and abandoning all illusion of control, I put my hands up to my face and screamed with all my heart, "Archangel Michael, help!" And then I hit the pillars of the bridge.

There was a blinding flash of electric blue light—and then nothing. I felt as if I were floating, suspended in a big blue bubble of protective energy so thick that no harm could befall me. Archangel Michael is the defender of the faithful and the protector of the innocent. I believe in the concept of Grace, and perhaps my earthly allotment of this precious quality was not yet used up. Whatever the reason, I found myself sitting in my still-running car, in the middle of nowhere on an icy stretch of bridge—completely unharmed!

Finally, after several minutes, I recovered enough to take stock of my situation. Trying to open the driver's side door, I discovered that it was tightly crumpled, so I had to roll down the window and climb out. Once outside the car, I was shocked to see that my whole front end was crushed up toward my windshield. I realized that it was the dead of winter on a deserted, snowy road. No one else appeared to have been so foolish as to drive in these conditions. If my car were not in running condition, I would probably perish anyway, as the wind chill had driven the temperature down to around 15 degrees below zero. I wondered if my life had been saved from certain death so that I could slowly freeze to death.

Resigned to deal with whatever came next, I climbed back through the car's window, slid behind the wheel, and put the car in reverse. I held my breath in fearful anticipation. The wheels spun a little bit and then found some traction on the slippery road. I backed up and threw the transmission into drive, continuing on my appointed rounds. I arrived without further incident at my destination and went to work.

When it was time for the return journey home, I pulled into a gas station and filled up the gas tank. Other than that, I could do little else to check the serviceability of my vehicle, as the hood of my car was so thoroughly crumpled and mangled that I doubted it would ever open again. Trusting that divine intervention was working well so far, I decided to stay the course and drove home to Bozeman, silently entreating the class of Angels who doubled as car mechanics to hold the car together just a little while longer. I pulled into my driveway at home, and just before I could turn off the key, the car's engine seized up and sputtered to a stop for the last time. The car was such a total wreck that I later had to have it towed away for scrap metal. Once again, my Guardian Angels had come through on my behalf, and gratitude doesn't begin to cover how I felt—and continue to feel!

⋰⋱

If you hit black ice on a bridge at 65 mph as I did, it creates a very interesting equation in consciousness. If your rule set for an encounter looks something like "a car traveling at 65 mph times black ice times bridge pylon, magnified by the force of the impact," then the output on the other side of the equal sign could add up to the probability that you might be dead or at least seriously injured.

If your equation has enough flexibility to allow for the presence of a hidden variable, such as angels, the balance of the equation is upset in the direction of less predictability. The real-world outcome could be a lot safer and more pleasant for you. The same equation with one variable added can make the difference between life and death, loved gained or lost, or success and failure. Try inserting the variable *faith in things hoped for and belief in things not seen* into your personal life equation as an experiment.

In my equation in the paragraphs above, the real numbers represent the speed of my vehicle, the iciness of the bridge, the loss of traction coefficient, and the force of impact. By the introduction of an imaginary integer into this equation, the imaginary or unseen number $i = Angel$, a different end result is obtained. The hidden variable could be said to represent the quality of faith in the unseen.

If believing in Angels is too big a quantum leap for you, don't accept that concept. Whatever you believe comprises the filters of your perceptions. How you choose to see things governs what shows up. For some of us, even the statement "I am the water of eternal life," spoken by the master Jeshua, could be taken to imply that God is always raining on your parade (especially if you live in Seattle!). Embrace the drop of mercy that becomes the ocean of life, love yourself, and transform your reality.

9
WAIT A MINUTE.
WHAT JUST HAPPENED HERE?

The *uncertainty principle*, formulated by Heisenberg, says you can't measure a quantum particle's momentum and its position simultaneously. If you track its position, you lose track of its velocity. If you track its velocity, you lose track of its position. My guides showed me something about that and why we cannot detect the particle's velocity and position simultaneously. What we observe with our physical senses is based upon the information filtered from our perceptions to our conscious mind. The left brain analyzes the raw data and then decides what information is relevant to our continued survival; all other data is suppressed. Though perceived or recorded, it does not pass the gatekeeper of our left brain and remains hidden in the dark recesses of our subconscious awareness.

Natural science does not simply describe and explain nature;
it is part of the interplay between nature and ourselves.
Werner Heisenberg

The conscious mind performs like a shutter; it takes very fast pictures and then places the information together seamlessly so you don't notice—just like movies, where the action is not actually a flow of movement but a series of static slides. If we are looking at an object or information, we can only consciously use less than 0.01 percent of the visible *electromagnetic spectrum* our eyes can detect. We cannot visually detect simultaneous channels of information because our left brain processes information in a "series" format, one slide at a time.

That is why we say that we can only process seven (plus or minus two) bits of data a second. The process we use to analyze sensory data works only in a linear fashion; it is very fast but only able to present one slide or slice of information at a time. Our unconscious mind is a *parallel processor* and can view millions of bits of data in the same second. Its conclusions are just not available to our normal conscious state.

That is where the measurement problem in quantum physics comes in. In order to analyze the data that our experiments generate, we have to use our conscious mind. Our left-brain awareness as a serial processor means that we can never see more than one perspective about the data at a given second. In addition, when we "look," we entangle the process of our observations with the object or subject of our perceptions.

Our conscious mind is like Jack Nicholson's character in *A Few Good Men*. Your conscious mind says "I want the truth!" You can almost hear Jack as your subconscious mind taunts, "You can't handle the truth!"[1] Jack is guarding the border, and in order to cross over, we have to change the side of the fence we are on. In order to perceive multidimensionally, we must use the right brain's parallel processor. However, in order to interpret the data we receive, we must analyze it from the perspective of the left brain.

The uncertainty principle only applies to our conscious inability to observe both an object's momentum and its position. Our subconscious mind can do it easily—while standing on the head of a pin and with an infinite number and variety of Angels! That is one of the best arguments for cultivating the ability to access altered states of consciousness. Since at the quantum level the act of observation changes the activity of the observed and you are composed of patterns of light, you can change your outcomes (and affect your income as well) by learning to perceive through the eyes of altered perception.

The effect of looking into the world of quantum physics is undeniable and drastic. It collapses possibility into actuality. *Unless you change your rule set, you get the actuality that there is no possibility that things will change.* This is because you have set up things in your life in such a way that they have to conform to the way you think they should. This is not at a conscious level. It is at an unconscious level. However, *it is consciously driven by the way you notice and perceive* in your physical reality.

By performing a measurement at the quantum level, we change what we are measuring. The act of observing at that level changes it. So if I notice that something feels "stuck" or "rigid," I am not making a judgment; I am making an observation. If I make an observation and then ask, "What would it be like if it were different?", I can notice that instead, and the outcome can be quite different. That question is open-ended—and open-ended process accesses the "*not* state of judgment."

In 1970 Lizzie James interviewed Jim Morrison of The Doors, discussing life in general and how he chose to live it. The interview was published ten years after his death in *Creem* magazine in an article titled "Ten Years Gone."

> *The most important kind of freedom is to be what you really are. You trade in your reality for a role. You trade in your sense for an act. You give up your ability to feel, and in exchange, put on a mask. There can't be any large-scale revolution until there's a personal revolution, on an individual level. It's got to happen inside first. You can take away a man's political freedom and you won't hurt him—unless you take away his freedom to feel. That can destroy him. . . . That kind of freedom can't be granted. Nobody can win it for you.[2]*

Your heart knows the difference between what is true and what is not. Your heart can judge because your heart is the only place where judgment makes any sense. That is where "ye judges righteous judgment" because you are not analyzing. When you are analyzing, you are always in a polarity or a duality with what you are trying to make sense of. It is either good or bad. It is either black or it is white. It is evil or it is absolute good. It is Angel or it is devil. It can never be both or neither—and that is the same attitude largely held in science.

COHERENCE AND DECOHERENCE AS SELECTIVE PERCEPTION

The concept of the collapse of the wave function assumes that beyond the domain of time and space there are infinite possibilities. Anything can occur. The laws of quantum probability imply that *the act of observing something causes it to cohere into what we expect to see.* When a specific outcome is chosen or observed, all the other possibilities simultaneously decohere in reference to this particular space-time domain. What we do not expect to see does not occur. There is essentially a precise formula that determines what is possible to manifest in your world. What you expect, you will experience. When you begin to loosen your rule set concerning what is probable, you will begin to see a lot of other things. Your "Reality Software" will receive a perceptual upgrade.

THE DOUBLE-SLIT EXPERIMENT

Physicists perform an experiment that by now many of us have heard of but bears repeating. They take two slits just wide enough for one electron to pass through and an electron beam gun, and they open one slit and fire the gun at a photographic emulsion on the far wall. When they calibrate the result from the photographic plate, they notice an equal distribution of particles, which is what they expected to observe as an outcome for the experiment.

However, when they open both slits, the electron appears to interfere with itself and go through both slits at once. Then it creates light and dark bands. What is that? Where the dark bands occur, it interfers with itself and cancells itself out. Where there are light bands, the waves multiply together, creating a summation potential (real physics) and it gets bigger, broader, and brighter. So the double-slit experiment is basically this: You have two slits, and you close them off. You have an electron beam that shoots one electron through the slit. Behind that you have a wall that has a device such that every time the particle hits the wall it goes "ding" and it also has an emulsion (a film or photographic plate) so that it records the pattern.

Well, here is the interesting thing: When you open one and observe it, it does exactly as you would expect. There is a very tight distribution of particle points. However, if you open both at the same time, something weird happens. The particle seems to split in two and interfere with itself, creating wave patterns on the film. When the scientists are just observing the experiment, it remains a particle.

If you are a parent, you know firsthand something about how this experiment can play out in real life. As soon as the physicists were not observing the path of the electron and both slits were open for the particle to choose, it chose to misbehave—just as your kids might. So you would get an interference pattern consisting of light and dark bands on the photographic plate.

John von Neumann, the famous mathematician, was brought in as an expert to try to explain the baffling results of the double-slit experiment. He determined that the presence of a hidden variable was the only possible rational explanation. According to Thomas J. McFarlane's paper "Quantum Physics, Depth Psychology, and Beyond," von Neumann went on to state that the x factor was human consciousness! He said that the reason that the photon or the electron interfered with itself is that our consciousness (that of the human being) actually caused the collapse of the wave function, which created the difference between whether it was viewed as a particle or a wave.

In other words, when we were not looking at it or expecting to see something, it behaved as if it were a wave and could interfere with itself. However, when we observed it and made measurements, expecting it to behave in a certain way, it did that. So based upon the idea that we are composed of photons, the act of measuring can cause the collapse of the wave function and change the actual structure of how a human being is put together. Essentially then, our universe is made up. Basically, consciousness is the x factor that is left out of all the experiments, but it explains most of the effects we see in quantum physics.

You can imagine complex patterns of photons interacting that together would compose a unique complex called a wave function. How your perceptual bias is set up defines the wave function of possibility that you can embrace and accept as an outcome. *The possibility wave is based upon your consciousness model.* What that means is that if you expand your model for your personal reality, your results change. They change exponentially.

These possibility waves are transcendent potentials, which are manifested by *freedom of choice.* You choose how things show up by the way you observe. Your observation or measurement creates the *collapse of the wave function.* Physicists are attached to this word to denote quantum measurement because of the image of spread-out waves suddenly collapsing to a localized particle; that's what it looks like.

When you stop trying to do something is when the change occurs. When you are trying to make something happen, you are in a particle-based reality with the information or object under consideration. When you go *whoosh* and expand your awareness from within the laboratory of your heart up and out, you are momentarily released from conscious limitations and transcend the barrier of the space-time domain.

In the next moment you will come crashing back from wavelike transcendence into particle-based reality. A portion of your consciousness, which is neither particle nor wave, can track both wave functions of probability and the particle-based reality simultaneously. And then the Angels and you can take up the tune and dance. Over time, the music becomes easier to perceive and, with practice, the dance becomes more fun!

When you collapse the wave function, it collapses from a state of smeared-out possibility into a state of definiteness. All you might need to do, if you have pain or a problem, is realize that your expectations are based on your perceptual bias. Your problem is always there, in part because you assume that it will be, so it shows up in the same way every time. In the next moment, it can be totally different. You assume it can't because physical reality appears to be unchanging. That is the rational assumption based on a closed-system model. It's a lie!

Your experience of personal reality is a completely open system. *Changing how you perceive changes the object of your observation, which can change your outcome right now!* It appears not to change because you have set it up in such a way that it has to conform to the way you think it should. This is not assessed or realized by you at a conscious level. It is consciously driven by the way you notice and perceive in your physical reality. Your habitual patterns of perception create the template from which you derive your daily experience.

Below is a very recent post from a seminar participant. Just by reading this book you are in the morphic field for the types of experiences described here. You do not really need to go to a seminar (although they are a lot of fun). I used this example because it was fun and instantaneous!

I went to my first seminar at the suggestion of my friend Suzanne. I had met her less than a month before and seriously debated going. Could I afford it? Was this stuff even possible? Ultimately I went, and I could not have made a better decision.

On Friday night, after seeing people fall over on stage and becoming somewhat concerned as to where they had "gone," I met Dr. Bartlett. He walked up to me and immediately identified my scoliosis. Within the next few seconds, I felt my spine leave my body—I'll say that again: my spine left my body—and come back completely different. After noting that some of my internal organs were jammed up in my rib cage, my lungs were not functioning at full capacity, and my hip was out of alignment, he told me to see how I felt. Still being hardwired to not really believe that such changes could take place in less than two minutes, I tried to walk away. That didn't work the way I had planned because my legs were now the consistency of jello.

After learning how to walk with a new spine (and staring in the mirror at what used to be a curved mess), I went home with the biggest smile ever seen. The next day I met a lot of great people, and I learned that a lot of things I've always suspected are actually true and that there are ways to quantify them. The day after that, I learned there's no need to quantify them . . . they just are.

So far, I've practiced Matrix Energetics on several people with encouraging results. The world is completely different to me, as I'm sure it is to many of you. Going to a seminar is the best thing you can do for yourself, your friends, and your family. If anyone is on the fence about going to a seminar, don't hesitate. GO! You will not be the same when you leave, and that is just fine!

—AM

10

COLLAPSING THE WAVE

When we observe, we collapse the electron's wave to localize at one place. Between our observations, the wave function spreads out into transcendent possibility. This is why I say *drop down* out of your head into your heart and enter an altered state. *Place intent*: "I'd like this to change!" This is really all the intent you need. When you access the feeling of a new outcome even for a moment, the feeling state of open expectation can be a quantum prayer.

Let go: Come out of space-time for a fraction of a second. In the next moment, you are instantly back, but your experience of reality can completely change. When you choose to observe your life from an expanded perspective, you collapse the wave patterns of your "same old, same old" existence into a new dimension of possibility. This creative act of Divine or Higher Will, combined with the limitless potential held within the space of your Sacred Heart, can encode and create a new you in the making!

THE FUNCTION OF WEIGHTED PROBABILITY

Trust the heart's wisdom. When you focus on what could happen, it becomes a function of weighted probability, which is generated and sustained by the way you have chosen to observe things. **Awareness, filtered through the medium of your beliefs, generates your experience.** No one sees the same event in exactly the same way. We are on the inside looking out. If you don't like how your life is playing out, let go of the need to control it. Drop into your heart center and trust whatever guidance you receive from within that point of balance.

For me, there is a key idea embodied within the following two quotes from the New Testament:

> *If a house is divided against itself, that house cannot stand.*
> **Mark 3:25**

> *No one can serve two masters. Either he will hate the one and love the other, or he will be devoted to the one and despise the other. You cannot serve both God and Money.*
> **Matthew 6:24**

Your beliefs about reality and about your life can enslave you to the kinds of experiences you are likely to have. When you focus on your pain, problem, or condition to the exclusion of all other possibilities, these issues become your master. If you wish to take advantage of the wisdom of the Master, you must decide what you choose to serve and to serve up as your reality. When you open up your expectation set to allow for a miracle, you are no longer enslaved by the structures of your previous creations. You can choose anew.

Realize that at the fundamental level, light makes up the house of your atoms, cells, and electrons. This is the composition of your physical body. When the focus of your attention allows only for the observation of your problems and the substance of the physical body, to the exclusion of all other possibilities, you are in the duality of service to two masters. In order to change your outcome or to have a miracle, simply let go for a moment. Allow the primal reality that is light to shine through. In so doing, you can redefine the structure of reality. Service of the light is the service of the infinite. Master Paramahansa Yogananda wrote about the teachings of Christ in *Autobiography of a Yogi*. In this book he describes his experience of the light of Oneness:

*. . . my physical body lost its grossness . . . I felt a floating sensation . . .
the weightless body shifted slightly . . . to the left and the right. I
looked around the room; the furniture and the walls were as usual,
but the mass of light had so multiplied that the ceiling was invisible.
"This is the cosmic motion picture mechanism." A Voice spoke . . .
"your form is nothing but light!"*

*I gazed at my arms and moved them back and forth, yet could
not feel their weight. . . . The cosmic stem of light, blossoming as my
body, seemed a divine reproduction of the light beams that stream out
of the projection booth in a cinema house. . . . As the illusion of a solid
body was completely dissipated . . . my realization deepened that the
essence of all objects is light.*[1]

In Matrix Energetics, we teach a mythology that states that we are
made up of photons. What we have done in physics is divide things up
into the realm of the very small. In this realm of virtual particles and
collapsing wave functions, reality behaves according to quantum rules,
which are very strange indeed. Then we say that large objects, such as
our physical bodies, are only subject to the laws of classical physics:
Newtonian-based laws.

From a scientific perspective, we are ultimately composed of atoms.
The atom breaks down into electrons, protons, and neutrons. Electrons do
not actually orbit around the nucleus as we were taught in school. Instead,
the electrons have what are called *probability orbits*. In essence, they only
have a defined orbit around the nucleus when we observe them. They exist
in an *electron cloud*. Then when we measure them, expecting them to be
in a particular orbit, they collapse into that orbit we expect to see. I am
still talking about you and what composes the apparent structure of your
physical body and surrounding environment.

From the threshold of the electron, we move deeper into smaller ter-
rains composed of virtual particles, such as muons and gluons, and so on
and so forth, down to the photon. The photon can either manifest as a
particle or a wave. (Or as Louis de Broglie would have it, as a particle *and*
a wave.) We now are dealing with light and information—and that is

what I believe we are made of. Not sugar and spice and everything nice, or snakes and snails and puppy-dog tails.

If, at a fundamental level, what we are composed of is a stream of photons held together by consciousness, then how are we not that at the macroscopic level of our bodies? So again, I remind you that when we measure something at the quantum level, we change it. It is called the *observer effect*. We cannot actually measure the velocity or the position of the quantum particle without changing it through the act of our observation. So the question to ask ourselves is, what lens of observation are we looking through?

11

COLLAPSING THE WAVE FUNCTION OF THE HUMAN EXPERIENCE

The language of the science behind Matrix Energetics' "two-point" process is useful. I believe that Matrix Energetics could not have gotten to its present level of sophistication without my growing love for physics and science. The two-point process provides a measuring tool, which supercharges our ability to notice where we can plug into the grid of the *All That Is*.

Now, how do you know what a measurement is? It is looking into someone's eye and noticing whether he or she is angry with you or excited or perhaps puzzled. You are performing a measurement based upon billions of calculations. All that complexity can be reduced to the simplicity of noticing what you notice and still remain completely complex. That is where you can pop into the grids of manifestation, without the need to understand any of it.

COLLAPSING THE WAVE

To remind you, two or more quantum systems can share the same quantum wave. When they do this, it can be said that they connect or become *entangled*. At the subatomic level, you are made of high-energy photons; your body consists of light and information held in patterns or waves of interference. When you connect the two points, you have consciously observed them as being linked. You have created that link with your imagination. What you imagine at the level of the photon has tremendous power to change these patterns of light and information. The act of focusing at this level, where everything is just made of light energy, causes what you observe to behave differently. You collapse the particle-based arrangement of your world into intricate patterns or wave fronts of light. Feel and sense this happening.

It is helpful to think that when you are doing the two-point procedure with someone else, you are in a very real way entangled with some aspect of yourself. Your experience of the other person is not the same as their experience of themselves, or even their experience of you. It is a uniquely blended state, and when you engage in the production of such an outcome, there is a unique opportunity for transformation of consciousness to occur. Through this process, not only do the things that you choose to focus on change, but you change and transform as well. By "doing nothing" and not trying to "fix" anything during this process, you are entering into transformation.

When you practice the art of Two-Point, it represents a new paradigm for things you can do or access with your sensory modality of touch. If you endeavor to do this on a daily basis, you begin to have glimpses of the hidden reality and its complexities behind the shroud of daily events. Things no longer happen to you. Instead, you begin to take responsibility for your creative use of universal energy.

EXCERPT ON THE TWO-POINT PROCEDURE

A Radio Interview with Dr. Richard Bartlett:[1]

Interviewer: How do you work with Matrix Energetics in your office, Dr. Bartlett?
Richard: I do not talk about health care because when you are talking about health care you are also talking about disease care. It is a balanced polarity of opposites, and any time you are working with one end of a polarity or a pole, you are working with another end. So if you are trying to heal some-

one, you are trying to heal him or her from a disease. Every time you do that, you can only use the parameters and the judgments that would be exercised in response to a condition.

Once you get out of the conditioned response, out of the tendency to react, a stimulus/response way of thinking, then what you can say is "I do not know what is going to happen next. I surrender to the moment and then embrace what shows up as being useful—and then learn from that."

How do you determine what you are going to do to help a person?
One of the many things we teach people in a seminar is to make distinctions by *noticing what they notice.* What that means in the way we are trained medically is that when someone comes into my office and says, "I have problems with my knees," and I look at their knees and fixate on their knee, I see them as having a knee with a problem. Then perhaps I add in a diagnosis that has already been superimposed upon that condition. I have now collapsed the probability of that pattern into a knee with a problem. That is all we are going to see, so at that point, we hope I am good with knee problems. That is the only option I really have left open.

At the other extreme, if someone comes in with a knee pain, I am certainly not going to respond by saying something like "I have a knee also." If I respond that way, then I deserve a knee in the groin. No, I will be respectful, and I will touch and hold their knee. It is kind of like a split or dual mind. I can go to the point of what I refer to in my first book as a "problem-set." This would represent the state of pain in the knee, replete with swelling and all the possible diagnosis and treatments, and whatever else has happened in the past. I can also *at the same time* hold an open frame of awareness for whatever shows up in the moment.

The way we are trained to focus on structure or matter, we look at the knee and our eyes focus on the knee. What if, instead, suddenly I am looking at the knee, and I target a space six inches out from the physical knee, and that is where my attention fixates? I am going to assume that the information six inches out from the knee is where I want to start to notice changes in that pattern. In other words, I will be drawn to an area and will focus on that, *but not to the exclusion of everything else.*

I will just allow myself to go into what is called *second attention*, let my awareness drift into what shows up, and then begin to work with that. It is like a conversation. Once you start to work with something like that, you interact with it and *notice what changes*. Completing that action, then you would notice where your attention is drawn next. If you do this and link them together, then what you do is create a specific syntax or language that allows you access to their problems, without entering into the problem state.

So it is really a case of going into a state of expanded awareness?

Absolutely. That is what we teach in the seminar. If we can get people into a state of feeling the change when they do something that we call the Two-Point, then when they feel it in themselves, and more importantly around themselves as expanded potential, they are now in a state where it really does not matter what they do; they will have an effect.

So what is the two-point technique?

Do you know how, when you take two poles of a magnet, like, say, refrigerator magnets, and you bring them together, at one point if the poles are matched they will push apart and if they are mismatched then they will pull together? You will get this dynamic tension in the air between the two magnets.

Touch one point on anything. It could be on a guitar or on a car or on a knee. Notice where it may feel stuck, hard, or rigid. It does not have to be painful. Rather, upon touching it, that point feels different from the rest of the area. That is the first point that you are going to choose to notice.

Now you have this first point. All you are doing is noticing something. Now take your other hand and just play around on your leg or any other object of your choosing in a different area—it does not matter where—until you find something that feels like it makes that first area harder, more stuck, more rigid, or like it is being pulled together like a magnet. That is what you are looking for. That is where you have actually performed the measurement aspect of what we teach with the Two-Point.

So is this a sensation that you experience in your hands?
No, you will feel this sensation as a physical hardening of the tissue or area underneath your hand. For example, you have chosen one point on a leg and it just feels a little different from the rest of the leg. Why does it feel different? It feels different because you are creating a game where you are choosing to look for something that feels different. It is not because it is different. It is because you choose.

Now as soon as you have done that, with the other hand, slide it along the leg or object which you have chosen until you almost feel a stick where you have a point that suddenly seems to draw those points together. That is an artificial, arbitrarily set-up measurement, but it allows you to play the virtual reality game. Now having connected those two points and realizing at the level of quantum physics that the act of measuring actually causes a change in your measurement, you let go as if you were dropping a pebble into the pond. (Do not let go of the object.) You imagine that you let go of the need for that to be physical, and you feel this expansive wave between those two points.

Many of you may experience a feeling of lightness, expansiveness, inability to think, joy, et cetera. Some people can experience other emotions, some people can see colors, pain disappears, all sorts of changes can happen. How is this possible? Realize that these two points no longer represent just a physical object. They are merely two points that you have chosen to measure. That leads you to all other points and all sixty-plus trillion cells and all the photons thereof and the emotions and the chakras. It links you to the experience of you as being human.

So what I perceive when I do that is a sensation of great warmth.
Yes, that makes sense. Warmth is one way to look at it. I want to share with you that if you just see it or experience it as warmth, then that is fine. That is an observable effect, and that is what we are looking for. But we also want to allow for the unobserved effects as well. In other words, if you do this and you feel warmth in your leg, you might also find that your relationship with your dog or your car or with the traffic on the interstate has improved.

You might find that your memory changes or diseases vanish or any number of things. Why? Well, because at this point, *you are playing by a different set of rules.* With those different rules, you are deciding to behave as a set of photons or as light and information. All you have to do to change a condition at that level is to change the information that is being given to that pattern.

Would you then suggest some sort of visualization of what you would like the pattern to look like?
No, because when you visualize something, then what you are doing is limiting the outcome to only what you can visualize. Really, what that does is it keeps you physical. It keeps you from collapsing the wave function of those photons into a new pattern. You see, it only takes the blink of an eye to collapse a wave function, and it might come right back to feeling the same. But what people tend to notice is there are either subtle or very strongly defined differences, and at that point, you have entered into a larger reality. Then it is a matter of practice. Remember, you are not with those two points trying to fix a problem. Rather, you are using this process as a metaphor for your life and for your entire conscious experience.

So you are viewing yourself as not just a pattern of photons but as a holographic pattern of photons?
That is correct. You are using the point you have chosen as a tool of measurement to measure any aspect of yourself. Now let us do this again. This time think about some nonphysical aspect of your being that attracts your attention. It can be your love life, your finances—whatever you are drawn to in the moment that might benefit from a solution. Now find a point on your leg that matches that feeling. Remember, it is not a leg anymore. It now represents your feeling about *x*, whatever that is.

Does it matter which hand I start with?
It does not matter which hand you use. It is not a polarity phenomenon. In fact, you do not need to use your hands. I am teaching your conscious

mind. So find a spot that represents your situation. Can you share what situation you have chosen?

It is a project.
Here is the point about projects. Make it more global than a project. Do not make it about a single outcome of the project. Make it about the feeling of the project. Now find a spot on your leg that matches that feeling. It is easier to do than to say because, really, you are making it up. You make up a spot and decide that is going to be it. That will be it. Now what you are going to do is find another spot. It does not have to be on your leg. It can be on your desk or on your microphone or your telephone. Any second point will do. The important point is to feel connected with those points.

Now, simply imagine that suddenly you let go of all your concerns, of all your affairs. You spiral up into the air as streams of light without a care in the world. Then it becomes quite different. Now, if you go back and remeasure that point with a thought about the thing you were thinking before, it is going to be different. Go back to where you were before and think about what you were feeling before. Feel how it feels. In most cases, you are not going to be able to define it in the same old way or it is going to feel very open or expanded or you may not be able to describe what you are feeling.

In my case it feels cooler, like a cool, refreshing drink on a hot day.
That happens to me a lot. I will feel like a cool wind went through my brain. It has a very physical feeling.

Also, I feel lighter and brighter.
Now notice how you feel. Take your attention off the point and notice how you feel. What do you notice?

Definitely a feeling of great positiveness, and optimism. It is a fabulous feeling.
Now expand that out from your physical body and notice how the space around you feels.

Now find a spot on your chair that attracts your attention. Do not make it about anything. It feels hard or magnetic or simply attracts your attention. Now find another spot, maybe on a part of your body, like on your shoulder or a part of your chest. Make that spot match the feeling of the chair. In other words, find those two spots and find the feeling of them as being related. It is really simple.

Now, do not let go of the spots. Let go of your consciousness and let it expand out into the universe. This is called the collapse of the wave function, and it does actually do that. If you touch your chair now, it is going to feel weird. It is going to feel expanded. It is very hard to describe, but people tell me this all the time at seminars.

Actually, my hands feel weird. I was actually not touching any point. I was about an inch away on both spots in midair.
That is great. You do not have to be touching anything. You can measure spots that do not exist because, in the quantum realm, none of that exists. You are making up your experience of the external universe and then building it up in your brain. Your eyes function as feature detectors, and you build up the holographic reference in your brain for what is out there. Some physicists who are particularly "out there" say there is nothing "out there" to measure.

What is interesting is my hands are still feeling very much as though I were manipulating subtle energies.
Realize that it is not about running energy, and it is not even all that subtle. Expand that and notice how your left big toe feels in relation to that experience. Until you are told to focus on it, you may not notice. But you will notice this effect taking place in your entire world.

How fascinating. Boy, that is an easy technique to learn!
It is so easy, it is literally child's play... playing like a child. The more you play like a child, the more expansive your awareness, the deeper your outcomes.

Children pick this up in a heartbeat. There was a famous university medical oncologist who was recently in my seminar. He came to the seminar very skeptical. I brought him up to the stage and did a Two-Point. He wound up unconscious, not having a clue as to what happened, and the whole room of one hundred and forty people felt a sense of sacred space.

You see, when you make this up and you make it up in a group space, it becomes like a sacred experience in nonordinary reality. It becomes very sacred and shamanic in tone, but also very physical, very reproducible, and very observable as a physical outcome. In other words, you can see scoliosis or a headache vanish. You can observe a jaw that does not move and suddenly the jaw moves instantly. Sometimes tumors or other physical maladies may vanish either instantly or over time. This is the thing. *You can only observe the things you can see. But changes you cannot see are also occurring, and they are probably the most significant changes.*

Time to go. So can you please give a one-sentence summation of Matrix Energetics?
Transformation occurs when you let go of the need for anything to happen.

TWO-POINT REVIEW

1. Locate a point on your body or a partner's that feels stuck, hard, or rigid when you touch it.
2. Find a second point that, when held in relation to the first point you are still touching, makes the relationship between points one and two feel even more taut, as if there is a magnetic attraction between the two areas.
3. Forming a somewhat arbitrary link between these two points allows for a measurement to be made. Remember that according to quantum theory, you cannot observe something without becoming entangled or interacting with it. The very act of observing the connection between these points with your feeling/imagination makes it so. This entangles the data and, in effect, collapses the wave of matter/consciousness that you have chosen to observe and to interact with.

4. Notice what is different now. The area between your two points probably feels softer and less rigid. You may notice changes in respiration; you or your practice partner may feel hot or flushed. It is not uncommon for the body to begin to sway or move to the beat of some unconscious primal rhythmic force. Stand behind your partner, because if he or she really entered the state I am describing, your partner may even momentarily lose consciousness. It is good to be prepared for anything, including spontaneous laughter, crying, or some other form of emotional/physical release.

KEY POINTS TO REMEMBER

1. It takes at least two points to measure anything.
2. In order to learn something new, you begin by noticing what is different.
3. Noticing what is different helps you suspend critical judgment and allows space for a new pathway of least action to be created. (In other words, you are creating a new activity that, with practice, becomes a new skill.)

Here is an example of how a first-time Matrix Energetics seminar participant corrected a painful condition through the simple application of focused intent on the problem using Two-Point:

I attended a conference at JFK in New York in late October. I was unsure of what, if anything, I could do following this experience, but I hoped to be able to help my wife with her shoulder, which had been troubling her for over six months (severe stiffness and pain). She was 80 percent cured with the first treatment, and the effect was instantaneous! Needless to say, she is thrilled with the result. That alone made the trip well worthwhile!

—*JS*

Basic Two-Point: Measuring like points on the body
(in this case, shoulders).

Checking relative motion in the shoulder points.
Noticing what I notice.

Basic Two-Point on the body.

Collapsing the wave. Obvious physical result.

Experiencing instantaneous transformation.

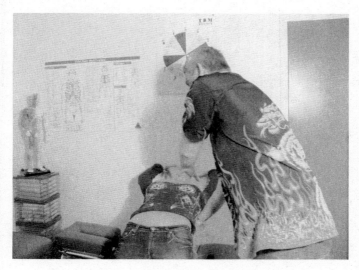

Experiencing instantaneous transformation
(continued).

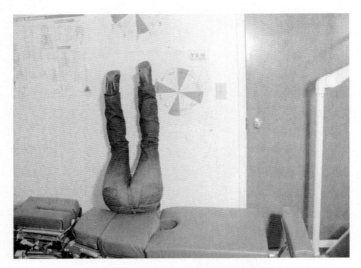

Her world has been turned upside down.

"Wow, that was a really big wave. Let's go again!"

Example of Two-Point off the body.

Moving into the Matrix state.

Experiencing a deeply altered state.

Basic Two-Point for a
shoulder issue.

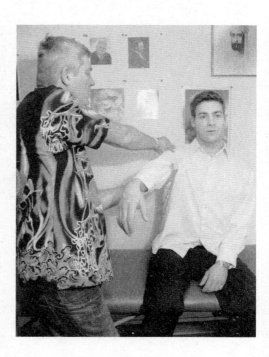

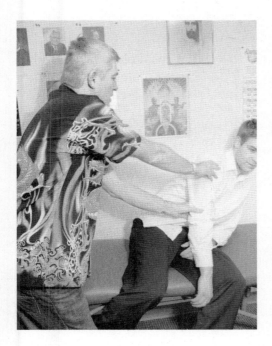

Feeling better.

Deeply relaxing into
heart-centered awareness.

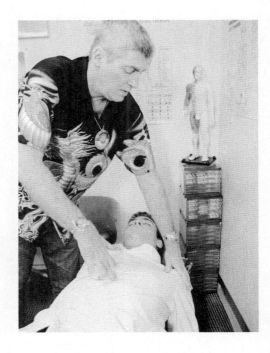

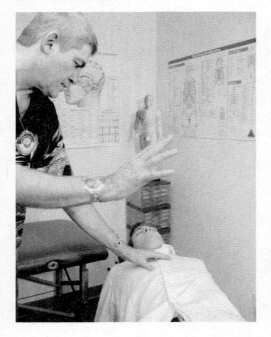

Two-Point working with expanding the field of the heart.

Two-point example on acupuncture mannequin for distance treatment.

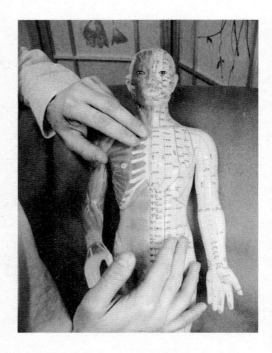

Two-Point on surrogate demonstrating distance treatment.

Two-Point for the energy field of the pelvis.

Feeling "stuff" change.

"Gone in
Sixty Seconds."

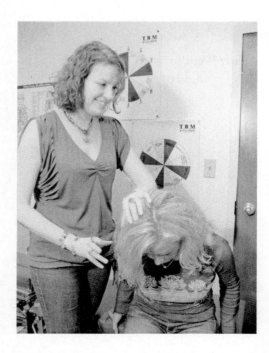

Just another day at the office (aka a seminar).

Working with the practitioner-certification
distance protocol.

12

TRUST WHATEVER SHOWS UP IN THE MOMENT

When we measure, most of us judge what we are measuring. We try to analyze it. We try to make sense of it. We try to make it mean something that fits our perception of the way the real world functions. There is something called innocent perception. Innocent perception or practicing the patterns of perception means that you notice whatever shows up. Our path of expectation assumes that there is a cause and effect to things: that if we do this, then this will happen, and if we do not, then something else will happen. There is no logic to it. But it is a logic we have embraced.

Your experience of reality, however you define it, is what you are able to observe and define. This does not mean that you cannot have an experience outside the confines of your reality. What it does mean is that in order to have an experience, you may have to step into a different way of observing, which is fine. One of the things you are going to learn in this book is how to *notice what you notice*. I am not referring to your ability to notice what you think you are supposed to observe or think. I will not tell you what the object of the game is but ask you to see with a fresh perspective how to notice what you notice.

If I look at you at a point right between your eyes, but my visual tracking drifts out several feet from your physical form, I am going to pay attention. Employing a soft gaze, I might ask myself the question, "What am I unconsciously noticing out there?" And then I am going to trust that information to be pertinent and useful to the moment. I am going to think that is worth trusting. So I go here, and way out there, and feel that. There. That gives us access to other possibilities.

If you find that your eyes go to the floor when looking at someone's chest, or they are drawn to your ear or someone's hairdo or a light fixture, then you trust that. When you do that it gives you access to other realities. Once you start to play that way, what has happened is you have encoded for a different experience. You are starting to see more of the information that is not normally accessible to your sensory apparatus until you assign the priority to pay attention.

All you need to do for massive change to be available to you is to notice something that gets your attention. It is about trusting and letting yourself go to the place where your attention is drawn. This is a measurement. Allow yourself to *notice something in the moment, and that will be your first point*. The measurement of the feeling or expression or anything that could give you a new reference can constitute *the second point*.

I do not know what that is going to be, and you do not need to know. This is a secret contract and is "eyes only" and on a "need-to-know basis." Just moving into a space where any aspect of your life can be different opens up a door of possibility for you to be there. You want to be in a space where things can show up differently instead of achieving the observed result of your normal expectations. Don't be overly concerned by what seemingly eludes your conscious attention.

It's okay to allow yourself to be perplexed or even to question whether anything I am saying is real or true. That is a healthy attitude to maintain, one of relaxed skepticism. Allow yourself to embrace the sense of not knowing, for that is where the power is. When you release the need to know or to do, you collapse the wave function of the problem-set into the unseen answer. Then you can precipitate within yourself the object or state of your desire. If your outcomes are always predicated on your conscious expectations, then where is the space you have set aside for miracles to occupy?

Dr. Hector Garcia, DC, is a good friend as well as a Certified Master Practitioner of Matrix Energetics and a certified instructor of the Yuen

Method. He uses his unique abilities as a medical intuitive to facilitate the different energy systems of the body. He talks about "noticing what you notice":

I notice things from within my mind and then project that awareness out into my external world environment. I am starting to do what Richard talks about by allowing myself to "notice what I notice" from "out there." As I have begun to utilize this principle, I have started to interact with whatever gets my attention or shows up.

For some reason, a lot of patients with tumors or growths come to see me. With a tumor patient, I will begin by calibrating the energy. By calibration, I mean I will touch the client on their body where the tumor is located, and this will become my first contact in my Two-Point. It is a point of measurement or observation. Next, I will look for a second point that gives me that stuck or rigid pulling type of sensation that Richard talks about in association with the phenomenon of the two-point process.

This second point does not have to be physical but can represent an energetic construct or concept. For instance, I want to know if I can calibrate the energy of the tumor in the client's body. Something like a tumor, though seen on an MRI or x-ray or other diagnostic tests, can, in my experience, have its actual causative energy pattern in one of three primary sources.

Locate where the Energy Is Coming From

*A tumor, or any condition for that matter, for my purposes, can be **of the body**, meaning that the tumor has a physical source and its energy is centered in the physical body. When I calibrate this, I know that if a diagnostic test is run in that case, there will often be clinically verifiable findings.*

*The second source of a pattern such as a tumor may be something that tests as coming **from the body**. In this situation, the tumor has come from within the internal bodily processes out into physical manifestation. This means to me that there is an internally caused source of the problem. This could be from **genetics** or a pattern in the **ancestral line** of the mother or the father's side of the family. It could result from a habitual pattern of belief. Or it could be a combination of genetic, emotional, and physical factors.*

*The third type of general pattern that I work with is a condition that comes **to the body** and is the causative seed of the condition. This type of pattern can be the result of a* *toxin or poison, something taken in from the external environment. It could be the result of an external trauma, which impacted the body, sometimes in the distant past, and set in motion the processes, which could ultimately lead to the formation of a tumor or other condition.*

In a more esoteric vein, a tumor or other disease pattern could result from taking in toxic belief patterns from another person or society in general—a belief, for instance, that a certain slice of the population is more statistically vulnerable to certain types of conditions or disorders due to factors such as race, sex, genetics, eating habits, mind-sets, cultural beliefs, or a host of other factors. These beliefs, or so-called scientific facts, can be in some cases the soil in which a tumor or other condition can be watered and nurtured.

IN THE BLINK OF AN EYE, YOU CAN BE MADE WHOLE

The quantum physics perspective can bring you to the point of oneness with the realm of magical possibilities. When you enter into the unified field of the heart, even for a brief blink of an eye, you can be made whole. This is why things such as diseases, cancers, phobias, and other conditions and limiting expectations can cease to exist when we do a Two-Point. When you cease assigning these things importance, they can cease to exist.

13

TIME FOR A CHANGE

In 1837 mathematician, astronomer, and physicist Sir William Hamilton suggested in his collective works that a science of pure time is possible, as it is well known that the fundamental units of time utilized in science are arbitrary. That means what? We made them up. It is even possible to construct all science on a single unit: time.

TIME TRAVEL AS AN EXPRESSION OF QUANTUM CONSCIOUSNESS

How do you measure time? You measure the scattering of *electromagnetic energy* over a distance, and distance times force equals your measurement of time. This shows the truth in Hamilton's statement. It is even odder because quantum mechanical time is not observable. It does not exist in this reality. Einstein said time is something clocks measure. He collapsed time into space-time. He said the two had to be collapsed together.

In physics, the mathematical term for time can be "plugged into" either the forward or the reverse direction. The equations work equally well either way. Whenever equations exhibit that kind of mirror symmetry, scientists tend to ignore it. They often say, "Well, that is just an artifact of the way it is examined." But what if it is not? There is a clue here that perhaps you can, in some unsuspected manner, move backward and forward in time. If the equations work in either direction, perhaps then you can theoretically find a way to take advantage of this to your benefit.

Photons can go backward and forward in time. A wave of photons traveling forward in time represents the "advanced wave," and the one traveling backward in time is the "retarded wave." Where these phase-conjugated waves of photons intersect creates the present moment. Why do you suppose we have a part of the brain called the temporal lobe? Fred Alan Wolf, an author and independent scholar and researcher in physics and consciousness, theorizes that it might have something to do with time travel. He has said that there is a time travel machine—and that it is our brain.

The present moment is now. What we call the linear experience of time is electromagnetic energy dissipating along a measured volume of space, according to Tom Bearden, a retired lieutenant colonel who helped design the Motionless Electromagnetic Generator (MEG) and continues to study scalar electromagnetics and unified theory.

We measure the distance traveled by the dissipating energy, and we call that the passage of time, or *entropy*. Time is how we measure work. It is what happens over time as energy dissipates. That is actually what work is.

Now what would happen if you flipped the polarity, or phase of time, so that energy is directed into the nucleus of the atom? If you do that, you get out the inverse pattern, which is negative energy or *negentropy*; it is *time reversed*. Negentropic energy doesn't dissipate; it coheres. If you reverse disease energy, you can engineer the cells back into their coherent and healthy condition.

When you generate a complex wave equation engineered by your intent and powered by your electromagnetic energy/feeling state, you are creating a closed loop that will return to you the answer to your request. This feeling state, or personalized electrical energy pattern, can serve as the impetus with which an *artificially generated quantum potential* may be created and sustained.

In the movie *Back to the Future*, the professor invented something called "the Flux Capacitor." Well, the Flux Capacitor may be based on an actual invention, the virtual flux of the vacuum, which according to Bearden is engineerable. When you curve space, it affects time. It changes it relatively. But there are negative space and negative time as well implied in this, and that energy may constitute the information in the vacuum. If you generate a unit of negative time, it actually coheres (goes backward), and this concentrates the energy rather than dissipates it.

The forces of entropy (measured in units of positive time, "actual" electromagnetic "EM" energy) and negentropy (measured as negative or backward flowing time, "virtual" or potential energy) are a paired manifestation. These paired forces represent the yin/yang of the universe. You can access that negative-space, negative-time energy and use it to curve local space-time. What is not possible within the constraints of classical physical laws can be circumvented.

When you apply the concepts of negentropy, or time-reversed waves, you begin to construct a scientific model for the reversal of disease conditions. If you also throw into this creative pot the potential to artificially construct templates or engines of consciousness with which to perform a specific task or function, the Physics of Miracles becomes understandable and reproducible. I am always being pushed to take things to the next level, whatever the next level looks like.

Collapsing the wave function of the problem-set allows for miracles to show up.

NEGATIVE TIME, NEGATIVE SPACE, AND VICTOR

My son Victor came down with chicken pox when he was about three years old. At the time, I had just listened to a talk by now-retired Lt. Col. Bearden. I did not understand a word of what he had said, but somehow it caught a snag in my awareness.

Something clicked within me and a certainty took hold. Thinking of my son's health, I then thought, "*Negative time, negative space.*" That was all I did. My only other thought was, "Go back to where you came from"— addressing the consciousness pattern of the chicken pox virus. Within the hour, Victor had no sign of his chicken pox. His fever and runny nose were gone. The spots were completely gone: it happened instantly! Scientifically, my explanation for what happened next is that I accessed the *scalar wave* potential of the *carrier wave* (more about scalar waves later), and it instantly inverted.

TIME TRAVEL

By utilizing the consciousness tools of Matrix Energetics, such as time travel, the seeds of many disease patterns can be reversed. You can "revisit" your childhood formative experiences without reliving or replaying the records of previous emotional charges and inherent traumas. You can learn how to change the wave patterns of your consciousness so that when you return to the present moment, whatever happened back in your childhood does not matter anymore. The possibilities are truly endless.

STEPS FOR EMPLOYING MATRIX ENERGETICS
TIME-TRAVEL TECHNIQUE

This is an exercise that is designed to take you through some steps to utilize the time-travel tool with another person.

1. Perform a measurement as previously taught in the two-point technique.
2. Ask the age of your partner or client. This will become your starting point, or as it's called in the movie *The Philadelphia Experiment*, your *zero-point reference time reference*.
3. Begin counting backward in five-year increments while holding your Two-Point.
4. Set your intent so that the quantum waves of change will activate when you "arrive" at the event, age, or time frame that you wish to interact with. You do not have to actually know the correct age of the event, because as you approach the approximate time reference, you will begin to feel the two points that you are holding start to soften and change beneath your hands.

5. Be prepared for the possibility that you or your practice subjects will experience physical or emotional releases of energy when this occurs. Gently support and comfort them, but try not to interfere with or edit their processing of the information and experiences.

6. When things have settled down and reached an apparent conclusion, reassess with your two-point procedure. Repeat the process if necessary, because there may be multiple time frames that need to be accessed to resolve the issue or pattern more completely.

TURNING BACK THE CLOCK

Let me share with you a story that clearly demonstrates the applicability of time travel as a consciousness technology. A woman came to one of our seminars, and when she returned home, she began a healing practice. She had a client see her in her clinic (her home), asking to be cured of a particular phobia. It was an odd request—a bird phobia. The woman, a newbie Matrix Energetics practitioner, didn't know what to do.

She did not know any phobia cures. So what did she do? She simply touched two points on the woman and conceptually "went backward in time" to before the lady had the phobia. The lady gently slumped to the floor, semi-unconscious, writhed around for a few minutes, and then came to. That was it. The next morning at 6:30, the fledgling practitioner (who had attended only one seminar) received a phone call from that client. The client was excited out of her mind. She said, "You will never guess what I am doing. I am calling you from my cell phone while I am hand-feeding the birds in my backyard."

14

MANY-WORLDS THEORY AND YOU

Einstein knew that there were not two separate sets of physical laws, although you can work with them and describe them that way. Einstein understood that the act of looking at something at the quantum level changes it, which is why he stated that if he *accepted* quantum theory, a mouse could look at him and change him. But Einstein and other scientists, being somewhat practical, said they did not see that happening. So why isn't that happening?

If we know that we are made of photons, we should be able to look at something and see it change, but just looking at Einstein from the mouse's perspective generated no apparent change in Einstein. To a physicist, if something is to be considered a law, then it must hold true all the time. So if it *should* change and it *did not* change, then there must be some other explanation. There must be a hidden variable—something we do not know or that the current theory does not allow into the equation.

Physicists Hugh Everett and John Wheeler came up with an explanation they named the *many-worlds theory*. This theory states that when you look at something, it splits into universes of infinite possibilities. The reason you did not see it (a mouse or whatever) change is because of the way you looked at it; your habitual way of perceiving caused your observed reality to remain consistent with your sensory-driven set of expectations. All the other choices not observed decohered from your reality and manifested, or cohered into existence, in some other parallel universe you are unable to observe.

How Choosing What You Notice
Plays a Role in What You Experience

In the realm unaccessible to your senses, whole new universes spring into being, branching off from the one you choose to adhere to, through the act of your perception. Things remain the same because you do not have the differential apparatus to see the other realities. In this realm, light is equated with the observer. Whatever is not observable in the electromagnetic spectrum, therefore, can be said to exist in a parallel universe.

The following story from one of my students is an excellent example of how parallel universes can be useful:

A couple of months ago, my friend Angela was told she had a major blockage in her heart. The doctors said she would need surgery, a stent, and a year or more of medication. I happened to read a blog she had written detailing everything that they told her was "wrong with her" and how afraid she was about the whole thing. Being in a state where I use Matrix Energetics every day, I decided to "check in" to see if there was anything I could "do." I instantly was able to see/feel the blockage in her heart. I was able to track the reality where she did indeed need the surgery and stent and medication. Knowing that was only one possible outcome, I shifted my awareness, and now I was witnessing the reality where the doctors went in and found nothing. I left it there, completely letting it go, and I didn't even tell her.

A couple of days passed and, lo and behold, Angela came back online with an update. "What the hell?" she wrote. The doctors had gone in and there was nothing there. No blockage, they didn't need to use the stent, and she wouldn't need to take the medication after all. Nobody could figure out what had happened; no one could explain it. Now, whether or not that healing might have happened anyway, we'll never know. All I know for sure is that there were a whole lot of people with pictures and lab test results who were convinced that there was the only one possible outcome.

Because of what I had learned in Matrix Energetics and my other studies about the nature of reality, I was convinced that for every situation there are many possible out-

comes. This is only one of the amazing experiences I have had since encountering Matrix Energetics. I use it daily, as it truly is more of a consciousness than a technique.

—*MB*

PARALLEL UNIVERSES AND TUMORS

My good friend Dr. Hector Garcia uses parallel universes to work with disease conditions such as tumors. He does his energy calibration where he asks, "Tumor of the body, for the body, or away from the body?" He calibrates or feels where the energy pattern of the tumor is located. He does not care if it shows up on a physical body. It may have its origin in other energetic realms, and what the body displays as a tumor in one way could be considered a reflection from the area of primary causation.

It is like the sun's reflection on placid water. If you have never seen the sun but you see its reflection, you think that it is real, rather than a reflection. In Matrix reality, whether what you notice is true or not does not matter. What matters is that somehow you get out of the way and you access something that makes a difference, a parallel universe or Bugs Bunny—it does not matter.

Dr. Garcia might say something like "You do not have cancer. It is in the fourth parallel universe. This parallel has cancer." So he sends the cancer to the fourth parallel. Now what happens if you send the cancer to the fourth parallel? Are you going to create cancer in a person there? No. If that person in the fourth parallel has cancer, and you send the cancer waveform signature to it, you are creating the anti-cancer wave. You cancel it out and it disappears from both sets of realities. If you cancel it out here, in this fourth dimensional space, the reflection is gone in the fourth parallel as well.

People used to worry that if you send it "out there," then you would give cancer to someone in another universe or dimension. No, you are sending out that force of negentropy that I described earlier. You are time-reversing the waveform of the electromagnetic signature of the cancer pattern.

As an example to help you understand this concept, pretend with me that you have lived in a dark cave all your life, as in Plato's cave allegory.

You have never seen the light of day directly, but you have seen the shadows cast on the walls of your cave from the sunlight. When you venture for the first time outside the shelter of your cave, the sun blinds you. You do not know what the heck that big shiny thing in the sky is; all you have seen your whole life is shadows, reflections of the primal cause or reality.

Now I want you to visualize a pond. The pond will represent, for the purpose of clarification, your physical environment. Your whole life you have perceived the world through the reflections on the surface of your own particular pond. What you are really doing is looking at the light of the sun as an indirect reflection on the surface of the pond. The light you perceive, in this case, a parallel universe, would be a reflection from a reality that does not exist in this realm; it represents a kind of virtual image or reality subset. However, given our perceptual bias and our ability to mold photons with thought, we make it exist. You realize that this reflection of light from the electromagnetic spectrum in a parallel universe is also a real thing; it is a *phase-conjugate*, a replica or mirror reality.

A mirror is phase-conjugated, meaning that if you look in a mirror, you see a reversed image. What you can do with a disease pattern is send it back to where the actual thing is (four parallel universes away). Now what happens to the disease, metaphorically, is that the reflection of the sun goes behind the clouds or is obscured, or perhaps in a parallel universe, the energy signature of the tumor pattern is actually obliterated in that moment. It disappears from the pond. Why does it disappear? Because you have phase-conjugated the information back to where it really originated, which, of course, cancels it out there. I think that may be one way Jesus healed withered limbs and other maladies. My whole life has been spent trying to get to this point of understanding the physics behind miracles.

PARALLEL UNIVERSES AND INTERDIMENSIONAL HEALING

When I two-point the energy, if I cannot find my second point on a client's body, then I go to where the information takes me. I often leave my hand on the area of concern, and I think "parallel universes," for example, and work with the concept as my primary point of reference.

Wherever I find it, I will "follow" by feeling or otherwise sensing the information. I will work with it wherever it shows up for me as the connecting link in my two-point process. Once I let go of the information, I allow whatever comes to me to happen as I drop down into an altered state of acceptance. I then simply get out of the way and allow the Matrix to reconfigure the pattern.

If my Two-Point calibrates (or "picks up") information for another dimension, this indicates to me that the information needed to change resides outside the normal dimensions of accepted space-time (such as length, width, height, and time) but still within what can be considered to be in this universe. I will follow the information wherever it takes me, noticing whatever I notice, and then let go of the need to "do anything" while universal intelligence and power restructures and transforms the pathological or distorted energy pattern.

PARALLEL UNIVERSES AND OTHER REALITY GAMES: DR. GARCIA

*If a person (or you) wants to work on a condition concerning eyesight, for example, I go to the body and feel for a connection in my Two-Point (which can just be in my mind once I get comfortable working in that manner). In this example, let's pretend that I do not feel anything there; nothing in the physical body energy grabs my attention concerning the expression of this problem. I **trust that** and look in a different way or I scan elsewhere. For example, in this case I don't "pick up" the information in this reality, but in a (perhaps a hypothetical or perhaps actual) **Parallel Universe**.*

Working with Dimensions and Parallel Universes
*In other words, when I say or perhaps just think the words "parallel universe," I feel a change in myself or in the other person or perhaps the field (space) between us. I sense an opening or a connection with the pattern of interest. That sense may feel different for you; the important thing is not how somebody else's process might function in a similar context. **Notice what you notice: whatever feels right to you. Follow the flow of information in a***

relaxed manner, wherever it takes you. Remember, one of the greatest keys to success with this process that I can give you is to stay out of your own way![1]

THE PARALLELS OF DR. DUNN

Dr. Mark Dunn (my student who eventually became my medical practice partner) was on a plane several years ago, and another plane came within an uncomfortably close proximity. When this occurred, something weird happened to Mark's reality. Nothing actually happened in this space-time, and the planes did not collide, but somehow Mark appeared to phase into what he came to think of as a parallel universe.

He arrived back in Seattle, and the signs in the airport looked vaguely different. Streets that he had been driving for many years did not take him to the old familiar places. The problem intensified when he went out to the pier near where he lived at the time and heard a strange, loud munching noise coming from somewhere under the pier. He asked several local fishermen if they heard that munching sound, but they looked at him as if he were crazy, carefully moving away from him.

The problem was accentuated when he returned to his office. The people who should have been on his schedule weren't. Individuals kept showing up for their regular appointment, but Mark had never met them before. To top off this case of weirdness, somehow my voice wound up in the background of his answering machine message. There was one little problem: I was not there when Mark recorded that message, so that was not possible! It took a few days but eventually Mark flipped back into this universe and everything stabilized relatively again. Yes, anything is possible—but the question we might ask ourselves is, *is it useful?*

SEE IF THERE IS A PARALLEL TO THESE EXPERIENCES SOMEWHERE IN YOUR LIFE

When we teach about parallel universes in the seminars, for some people it is a metaphor. For a few, it is a "met-a-floor" event, and for the occasional lucky one, it becomes a way of life. Mark just recently told me that he feels he has gathered enough awareness that he might actually be able to step into a parallel universe, or place, and fully inhabit it. I told him if he found himself going through a set of doors presented to him and above one was a sign that read "Gates of Hell" and above the other "Gates of Bill," choose the latter. "Good point," he replied.

You do not want to just go willy-nilly wherever your fantasies or visions take you. Trust your inner guidance and consider installing a virtual reality GPS. I wonder how many of us, who have lost our way in Life, could benefit from installing something like that within our consciousness.

Resonance with Concomitant Self

I read a fascinating science-fiction novel titled *Resonance* by Chris Dolley. In this book, the central character's timeline is becoming unraveled. Without warning the man would find himself within a different parallel reality where some of the details, such as where he lived or where he worked, had changed. As the story goes on, we find out that there are hundreds, perhaps thousands of versions of the central character. These parallels are beginning to converge due to a resonance that is set up as a result of similar choices and actions across the grid of multiple worlds. It is a useful concept and probably not an entirely fictional construct.

An oscillating, tapestry-like pattern or grid can become woven into the structure of your personal space/time. The presence of this increasingly powerful grid can form a strange attractor, which can allow for the increase of synchronicities or so-called lucky happenstances in your life. This is an interesting and useful idea, because if you are not living the ideal life that you would prefer, inevitably some parallel "you" is!

Resonance with Our Selves

You can *begin to resonate unconsciously with the right choices and lucky synchronicities that your more successful counterparts enjoy.* You can begin to bridge the timelines and merge them in the here and now. You can, in effect, create new opportunities, skills, and events in your life. Over time, this practice of parallel mindfulness will steer you to the convergence point where you begin to live the life of your grandest dreams. All you need is a little help from your parallel-self friends. If this sounds crazy, be advised that the most successful military remote viewer, Joe McMoneagle, believes that his success at remote viewing is due in part to information that another self sends him from twenty-five years in the future!

15

THE PHYSICS OF MIRACLES

In the 1800s, James Clerk Maxwell, a Scottish mathematician and theo-retical physicist, developed the rules for how electricity and magnetism behaved. He combined electricity, magnetism, and gravity in quaternion algebra. There were twenty equations to this theorem, which collectively described very succinctly the unification of electromagnetism and gravity. It was the original unified field theory.

When Maxwell died, Oliver Heaviside, who had worked with him, apparently decided that quaternion algebra was a headache. Heaviside got a little bit heavy with his editing finger and he took the quaternions out of the equations and substituted vectors. When Heaviside took out the scalar terms (real numbers and real parts) in Maxwell's original quaternion equations, we lost the ability to unite electromagnetism and gravity. We lost the *scalar potential*.

Einstein studied the truncated equations and concluded that you couldn't bend local space-time. The principles of relativity say that you cannot have a local curvature of space-time. If you can't curve local space-time then you can't transcend gravity because it takes a large body, such as the sun, to curve space-time. Einstein was working from faulty assumptions when he developed his theory of relativity. The quaternion math yielded the normal electromagnetic field and a *scalar field*, which when combined yields the potential for a *unified field theory*. That is right. At the time of the Civil War, Maxwell was able to unite the forces of electromagnetism and gravity through his quaternion approach. Unfortunately, when he died the original equations were lost and Einstein learned the truncated versions.

It was Heaviside's *vectors* that electricians studied in school, not Maxwell's quaternions. It's really hard to understand how Heaviside seemingly just said to himself, "Let's just get rid of that." We lost the understanding of Maxwell's *unified field theory* right then and there. The simpler vector-based approach of what was to become the current version of Maxwell's theory of *electromagnetism* lost something important in translation.

A scalar potential defines two vectors that, when added together, add up to zero. The two vectors' angles are perpendicular to each other and they sum up to a "directional vector of zero." The two perpendicular waves cancel out the directional component. When you have such a situation, your combined vector has no defined direction that can be measured; only a force of magnitude remains. Since no internal geometries are even posited as the result of different multivector systems, the net sum of a scalar vector would be zero. The geometries that the vectors can be configured with will change the *quantum/information potential* of the scalar wave.

This potential can be projected into the vacuum, creating a virtual stress into the nuclei of the atoms. Once this stress pattern builds sufficient momentum, it can be linked to a torsion field, which can travel far in excess of the speed of light. In the vacuum there continually exists at any and every point, and in any and every region, the ghostly image of anything and everything—whether in the past, present, or future and whether potential, probabilistic, or actual.

The universal vacuum may be taken to be a sort of giant hologram. The virtual replicas of all matter, including living organisms, exist in the vacuum as virtual or ghost forms. These are constantly shifting, changing, and transforming in and out of existence. You could have a particle, wave, *antiparticle*, or *antiwave*. Even an Angel could pop out of the void, becoming temporarily solid.

This hyperdimensional being could readily move from holographic virtual dimensions into actual physical reality. This Angel pattern might pop into our four-dimensional space, perhaps just long enough to save a falling baby or even prevent you from having an accident. With the emergency safely over, that same Angel might pop back into the hyper-dimensional realms, the spaces between the spaces, until needed or called into being again.

What this means is this: *"Real stuff" shows up when enough "virtual stuff" becomes adequately activated by our thought and feeling projections.* Then, over time, what we have created with our willful intent diffuses and dissipates from this reality. This occurs in part because we have not developed the ability to hold it in manifestation by loving it enough to sustain its continued existence.

If you start with the principles of scalar electromagnetics and com-bine that with the developing physics of *torsion* and *spin*, you can embrace elements of both as a starting point. *Now we are starting to unlock the secrets of a physics that is powered by the force of conscious-ness.* This new approach may provide profound keys to some of the alchemical secrets that have been sought since the beginning of time. By merging scientific theory with spiritual technologies, *we can begin to create a workable, unified field theory for consciousness technology*.

THE POWER OF MIND OVER MATTER

Scalar electromagnetics is partly a consciousness technology, which is what makes it so useful for Matrix Energetics. You can also utilize phase-conjugate technology to reverse the course of a disease. I believe that the Matrix Energetics time-travel technique may work according to this prin-ciple. You can utilize the two-point and time-travel methodologies in order to perform a type of mental targeting. This mental projection of scalar waves linked with specific advanced technologies is called *psycho-tronics* in Russia and *radionics* in the West.

Albert Abrams, MD, is the father of *radionics*. He noticed that when he tapped on someone's abdomen, he could elicit different sounds or the sound of a tympani, depending on what condition or disease the patient had. One day in his medical practice, he accidentally noticed that whenever he moved a pathological tissue culture near the abdomen, it changed the sound given off when he percussed the patient's abdominal quadrants. Abrams paid attention to what he noticed and asked new questions of himself. From there he made up radionics.

What he then noticed was that these pathological cells gave off information. For the next step in Abrams's experiments, he attached the dish containing the pathological tissue culture to a long, thin copper wire. Abrams would then touch the distal end of the wire to the patient's abdomen. When the doctor percussed the abdomen, depending on what culture was in circuit, the timbre of the sound varied significantly. Abrams was actually able to correctly diagnose the patient's disease by determining which pathological culture elicited the greatest change in the abdominal sound when percussed!

As time went by, Abrams realized that in order for the diseased tissue culture to be able to make changes in the abdominal sounds, he must have been dealing with some kind of subtle energy, since the patient had no physical contact with the pathogen. Abrams then hit on the brilliant idea that the disease or condition could be represented by a specific set of electromagnetic frequencies. A true medical pioneer, Abrams began to transmit the electromagnetic signature of the pathogen down the copper wire to the patient. At this point a strange and wonderful thing happened: Sometimes the patient was cured of his or her disease or condition!

With further experimentation, Abrams developed a machine that was designed to duplicate the frequency signature—what we could call a complex carrier modulated EM waveform. Abrams discovered that if he had a sample of the patient's hair, blood, or saliva, he could determine the diagnosis of the patient and treat him or her with just the EM waveform that would arrest the progression of the disease. Abrams abandoned his traditional medical practice and began to manufacture the strange "black boxes" for other practitioners to use.

The path-Oscilloclast was Abrams's crowning technological achievement. It looked like a big Moog synthesizer (for you Keith Emerson fans) with a bunch of wires and row upon row of knobs that were used to tune in to and isolate the specific frequencies to be used in the cure. One particularly enthusiastic owner of the path-Oscilloclast was curing many disease conditions, including cases of cancer. This doctor was excited by the results that he was able to get with Abrams's radionics technology. Then one day he looked down and realized that the machine had never been plugged in!

The machine hadn't been doing anything at all; the subconscious mind of the practitioner was the agent of the healing force. The patterns of your thoughts represent information stored as complex electrical potentials. You can imprint your thought forms onto a scalar wave and deliver the product at a distance.

It is my contention that you can learn to consciously direct your focused intent into the energy of the vacuum. *You can structure it and create an engine of consciousness that sustains it.* Your amplified creative thought activity builds and sustains a unique morphic field. This template or grid of your creative intent can be imprinted with the mechanism or design of your choosing.

Human intent can be psychically directed into the medium of the vacuum. This thought pattern can then draw more energy unto itself from the torsion fields present there, thus creating what Bearden called in his book *Excalibur Briefing* an "artificial active potential." This focused intent can occur in the blink of an eye and then be released to the parallel processor of the subconscious mind. If you do this with enough subtlety and with a certain cultivated sense of nonattachment, you perform a weak quantum measurement. This weak or egoless thought wave doesn't cause a collapse of the quantum wave function.

I think that perhaps our DNA is an antenna that picks up information from our environment, including our beliefs and our emotions. It then conveys that quantum potential to our bodies. When we unfold the imprinted information into our *bioplasmic field*, we then organize it into our bodies. We can literally manifest diseases in our bodies or not based upon what we are congruent or not congruent with. To say this in another way, whatever we suppress will imprint into our morphic field and be impressed onto our biology. That is why we really have to watch what we think.

In theory, you could simply reverse-engineer, or phase-conjugate, the disease pattern. You take it out of phase and then you have the healing pattern for the specific disease, which is just an electromagnetic wave signature. You can then "dial in" the harmonic signature or pattern of the distant target and then pop it back into the electromagnetic spectrum at the target.

In order to affect a healing, the antenna of the DNA would pick up the quantum potential for healing from the *zero-point field*. You would decipher it at the subconscious level as a bidirectional EM energy signature. The deciphered information would then be integrated into the level of your biophotons, creating greater coherence. Theoretically, the disease pattern could then be reversed. When you do this, you start to bridge the gap between a mere conceptual framework and a workable or engineerable outcome. If you do this, you will create a shared reality where it is more likely that you will be able to do some of these things, such as instantaneous healings and other miracles.

I think all our healing models are wrong, or at the very least incomplete. They are based on the standard models of physics that do not get you to miraculous instances. They do not increase the likelihood that you will access miracles. However, this physics does. This physics says that not only are miracles likely but they are probable. If you start to explain this to people, then you are taking it down to a level that is applicable to the average person in everyday life.

If you do this consistently, you will become more consciously aware of your ability to direct healing energies in this manner. With continued practice, you will begin to function from the awareness of your heart field, which is a unique biodynamic torsion field. When you do this, you begin to gain access to hyperdimensional states of reality. *This explanation helps define and make accessible to anyone the mechanisms behind miracles.*

A seminar participant recently sent this healing story experience to the Matrix Energetics message board:

I was the redheaded lady in the blue-green knit top whom you called up when you asked for people with gut problems at the Denver Level 3 seminar. Since you didn't ask me what was wrong, I wanted to share.

At the time you asked for gut problems, I was in an active IBS attack complete with spasms and chronic discomfort. I have had gut problems since I was born (colic eight hours a day for six months), but didn't begin the full-fledged IBS until after an ovarian cyst removal twenty years ago this year. Eating out, particularly on trips, has been really difficult because it's been hard to stick to exactly what will keep my gut happy (or at least merely grumpy).

When you attached that (imaginary) copper line, connected it to the "make-believe" radionics device and turned it on . . . I just felt a whole-body field permeate me and then, boom . . . I was on the floor. When I could finally get back up and go back to my seat, I felt my gut relax for the first time in two decades. I could literally feel the inflammation dissipating. It was amazing. I'm not sure my gut has ever felt that calm.

I tried it out on Saturday just to see what would happen. We stayed an extra day and went up to Echo Lake and Mt. Evans. There's a lodge by the lake that sells buffalo chili. Yup! I tried it with no ill effects. Yum!

—CP

16

THE SCIENCE OF INVISIBILITY

It was a sunny summer day in Seattle, June 1989. I was in my third year of an intense four-year program at Bastyr. At least this year I was allowed to see patients as an intern at the student clinic. My fifteen years as a chiropractor definitely helped when it came to my clinical skills set. One thing that Bastyr did not teach, and you had to learn for yourself, was the fine art of rapport. It was something that over the years in practice you mastered—if not your business failed. I learned early in my medical career that if people tended to like you as a person, they were more likely to trust you as their doctor.

That morning I had seen four patients on my shift at the university clinic. One had allergies and wanted advice on her diet. The second patient was trying to lose weight and had a history of adult onset diabetes. She was scheduled for some blood tests so that I, in tandem with the licensed naturopathic physician overseeing my shift, could formulate a treatment plan. The third patient was a follow-up visit for lower back pain that had greatly improved since his last visit. The fourth patient was brought in by his mother for advice on the possible links between his so-called ADD and his typical junk food diet. Finally finished, all my charting done, I left Bastyr for my afternoon private practice as a state licensed chiropractor.

Chiropractic care: is that what I was doing now? Frankly, I didn't know what to call what I did anymore. It would still be another five years before I realized that this weird phenomenon that occurred when I touched people was actually reproducible and teachable. Since that event, detailed in my first book, things had never been the same. These days it was rare for me to have to physically adjust someone anymore. The simple act of touching them lightly would cause their spine to shift to a healthier position, and their bones and muscles would realign themselves. Talk about an easy way to make a living! Thank goodness for that, I thought to myself. School was so intense that I sometimes thought the experience would be the death of me. It was a real joy to go to my private practice and watch what sometimes looked and felt like miracles occurring.

I had just finished lunch and was eagerly awaiting the arrival of my two o'clock appointment, a new patient. Two o'clock rolled around and then 2:15. It looked like she was going to be a no-show. No matter; it was such a beautiful day that I relished the thought of some peaceful time off from my hectic schedule. I decided to take advantage of the sunshine and went outside to lie on the hood of my '66 GTO, basking myself in the warm rays of the sun. Soon I was drifting off into dreamland. Suddenly a bright light intruded on my inner awareness. No, it wasn't that the sun had just reemerged from behind a cloud. This was something else entirely!

A bright form in the shape of a man appeared to my inner sight. Apparently this apparition had not a moment to waste as it resolutely informed me that in order to heal others, I should "treat with counter-rotating fields" and "study phase-conjugation." No sooner had this been said than the Angel—or whatever it was—disappeared, leaving me in the dark! It has taken me nearly ten years to piece together the beginnings of an explanation for what was so quickly and mysteriously imparted to me on that day.

Counter-rotating fields? What could that possibly mean? Then I remembered my fascination with the Philadelphia Experiment when I was much younger. Could that provide a clue to what the Angel meant? And phase conjugation; what the heck is that? Wait a minute; there was that guy, Tom Bearden, who talked about scalar electromagnetics in the 1980s at my spiritual community in Montana. His talk was about the weaponization of some strange concepts in a physics I had never heard of—scalar physics, wasn't that it? Didn't he say something about phase-conjugation and how this stuff could be used to heal disease as well?

I got all excited about the idea and resolved to pursue it—the best-laid plans of mice and medical students! Taking an average of thirty-one credit hours plus time spent in the student clinic left no time for anything but sleep. A medical student is more or less in permanent survival mode, not exactly a conducive mind-set for learning anything else outside of the rigidly defined curriculum. It was not until last year that my life and schedule stabilized enough to fully delve into the mysteries the Angel had imparted. What I gradually found out became the basis for much of the material in this book you now hold in your hands.

I believe this overlooked science of scalar electromagnetics is what the shining figure on that sunny, fateful day was referring to. The Philadelphia Experiment, even though perhaps more legend than fact, provides a powerful metaphor, as well as morphic field, from which many useful Matrix Energetics concepts can be derived. Read on and I think you will understand why this is an important topic for this book and for you!

When I was about twelve years old, I remember reading in my local news-paper, the *Daily Oklahoman*, about the Philadelphia Experiment. The funny thing is that if it occurred at all, and I tend to believe that it did, the experiment occurred in 1943! It makes little sense that an article about it would be in the morning newspaper in the 1960s. But that's what I clearly remember.

When I read the newspaper article, there were just enough details and talk about invisibility—and vague rumors of secret government experiments—that I resolved to find out more about this mysterious sub-ject. I went to my local library and checked out the book *The Philadelphia Experiment: Project Invisibility* by William Moore and Charles Berlitz. In their book, Nikola Tesla was named as one of the scientists who worked on the Philadelphia Experiment, so I returned to my local library to read about him as well. I cannot remember the title of the book I read, but I was absolutely fascinated by his life story, which left an indelible imprint on my young and very curious mind.

I also had to go back and read everything I could about this eccentric genius in pursuit of my quest to understand what I had been given that day as I lay on the hood of my car. For the last ten years, I had searched to begin to put together the very important pieces to the puzzle I had been given. Little did I know then that some forty years after reading that intriguing newspaper article, my research concerning the Angel's message would lead me right back to the very same books in search of answers!

WHAT WAS THE PHILADELPHIA EXPERIMENT?

The "Philadelphia Experiment" is the name that has commonly been given to an alleged top-secret experiment conducted by the United States Navy in 1943 in which the destroyer escort, USS Eldridge, outfitted with several tons of specialized electronics equipment capable of creating a tremendous pulsating magnetic field around itself, was first made invisible and then transported, in a matter of moments, from the Philadelphia Navy Yard to the Norfolk Docks and back again, a total distance of over 400 miles (640 kilometers).[1]

According to at least two of the many and varied reports, a nearby merchant ship crew purportedly witnessed the destroyer's arrival in Norfolk, via teleportation, and its subsequent disappearance, although later investigations and research could not substantiate the witnesses' claims.

Morris K. Jessup, a bootstrap astronomer and ufologist whose involvement in the experiment was in itself quite mysterious, claims the operation was a secret experiment conducted by U.S. Navy to "test out the effects of a strong magnetic field on a manned surface craft. This was to be accomplished by means of magnetic generators (degaussers)." In other words, the object was to render the navy ship seemingly invisible so that it could get close enough to other (enemy) craft or explosive devices to destroy them before being destroyed. The USS *Eldridge's* "teleportation" may actually have been the accidental "result of this trial invisibility run, involving a related time-warp phenomena," according to Jessup.[2]

The experiment seemingly generated a hazy, green, luminous mist (similar to reports of the Bermuda Triangle fog) that enveloped the entire ship: both the ship and its personnel began to disappear from sight, leaving only the destroyer's waterline visible.

THE "REAL" PHILADELPHIA EXPERIMENT?

Bob Beckwith is an innovator who has patented many electrical systems over his long career. In 1942 he invented a device called the "frequency-shift keyed transfer trip equipment." This technology, Beckwith claims, was utilized by the navy in an attempt to defeat a new type of German mine. It was hoped that Beckwith's device would help enable detection of the German mines at a safe enough distance to avoid or detonate them safely, and it was this technology of his that was the prelude to the alleged Philadelphia Experiment. So far, this seems somewhat congruent with Jessup's version.

However, Beckwith believes that the *real* Philadelphia Experiment was initially conducted in the Long Island Sound, utilizing an experimental minesweeper called IX-97. The description of this experiment also duplicates many classic elements of the Philadelphia Experiment legend. Beckwith believes that the lack of traceable or verifiable information associated with the Philadelphia Experiment is an example of deliberate obfuscation and disinformation by the Office of Naval Research.

According to Beckwith, three special-looking generators were driven by the ship's power. Controls for these generators were housed in a second cabin at the stern of the ship. Beckwith believed that three-phase currents were placed through the wires of the generators at a very low frequency, most likely at 7.83Hz—the Schumann or so-called earth-resonance frequency. These time-travel generators consisted of three single-phase units placed 120 electrical degrees apart. Each unit was approximately five feet tall by two feet in diameter. The generators put out a low voltage but emitted more than a thousand amps.[3]

Beckwith's *real* Philadelphia Experiment aboard the IX-97 was actually a time-travel teleportation by Edward Teller—a refinement and elaboration on Nikola Tesla's 1907 experiment where he allegedly moved an object along a laboratory bench. Tesla then turned on his electrical device and the object moved back in time to its original position.[4]

Teleportation? Invisibility? Time travel? What does all this have to do with Matrix Energetics? You can step into the field of consciousness that Matrix Energetics has built because we have said, "What if there were no rules?" What if we had a rule that said there are no rules? The Philadelphia Experiment has not been substantiated in any official manner; however, the story itself and its nature suggests that "the rules" we assume to be operating and in place may not be all there is to know—or all that we already know. That there are no physical laws is essentially what yogi and spiritual philosopher Sri Aurobindo taught. He said they were more like suggestions. I think it's perfectly okay to get a little bit fuzzy about the physics. **If the laws of physics are not really laws, but more like suggestions, it is far more likely that you will witness a miracle.**

There is at least suggestive evidence that we have technologies that operate from principles that utilize electrogravitics or antigravity, just to consider two examples. If this is true, I believe that consciousness comes first and technology follows. If we have material technologies that can do these things, we can do them as individuals as well. We inherently possess the *spiritual technology* for levitation, invisibility, and miracles of all shapes and sizes. ***This technology of consciousness resides right within the torsion field of our heart and is linked to our bioplasmic energy fields***. I believe it is the physics Jesus utilized. It is the physics of Tesla as well. So that is what we are beginning to teach with Matrix Energetics. The cool thing is that not too many people are teaching this. So what do we have? We have an unpolluted morphic field that we can build however we choose. Why not allow for the possibility of invisibility or time travel—or any other useful miracle?

DIVIDING SPACE

In her book *The Philadelphia Experiment Murder*, Alexandra Bruce examines the fascinating book *Hypotheses* by Beckwith. Among the other intriguing concepts Bruce covers, she notes that Beckwith's book develops a complex perspective that offers readers insight into the actual *physics of miracles*. Bruce explains that all atoms in universal space are energetically connected by what are called "strong nuclear force lines," which literally hold the universe together and function as a medium for the transmission of all frequencies. She states that these lines of strong nuclear force "can be broken with the application of what is called 'a three-phase neutrino field,' which in turn creates a bubble of 'divided space.'"[5]

Beckwith further explains:

A space divided from universal space can be created by causing a small percentage of neutrinos [that permeate all matter in our Universe] passing through the space to travel in a vortex rotating at a frequency in the order of 7.5 Hertz. Strong force lines at the boundaries of the space are interrupted so long as the vortex exists. This twisting field is necessary to break the field of strong force lines between all matter in universal space and to create an inner space separated from universal space.

If a rotating magnetic field [a key principle in the fundamental operation of the alternating-current—electric—motor] operated in synchronism with the Earth's 7.32-Hz fundamental resonance, objects within one such space can be moved with respect to our "universal space" when power is applied. The divided space is then free of forces of inertia or gravity. Once the space is divided, objects within the space may levitate, teleport, or move in time. Divided inner space can pass through universal space but is dependent on the drag and surface sharpness between spaces being low enough to prevent piercing the shell of the missing strong force lines. Electromagnetic waves—including visible light and infrared heat—can pass through the boundaries of the divided spaces.[6]

The book *Secrets of the Unified Field* by physicist theologian Joseph P. Farrell describes how the Philadelphia Experiment was probably done with torsion fields. Torsion fields create a hyperdimensional geometry, which has access to extradimensional realities. When you master the unified field of the heart, you can actually fold space-time locally. If you do that, then you can theoretically disappear from amid your enemies or reappear somewhere else.

METAMATERIALS AND INVISIBILITY: FACT OR FICTION?

James Bond's Aston Martin V12 Vanquish in the movie *Die Another Day* was able to activate an invisible mode by projecting images photographed by tiny cameras on the car that were then projected onto the car's light-emitting polymer skin. A fantastic notion? Read on.

> *In an example of life imitating art, professor Naoki Kawakami of the Tachi Laboratory at the University of Tokyo [along with two other professors] has developed a way to render a person partially invisible by photographing the scenery behind a person and then projecting that background image directly onto the person's clothes or onto a screen in front of him. As seen from the front, it appears as if the person has become transparent, that light has somehow passed right through the person's body. This process is called "optical camouflage."*[7]

Admittedly, this is a far cry from the invisibility cloak worn by Harry Potter. Author Syed Alam interviewed professor Susumu Tachi, another of the three professors now famous for their 2003 optical camouflage breakthrough, and gives us more insight into how this works:

> *In reality, the optical camouflage cloak is anything but invisible. It is made up of "retro-reflective material" coated with tiny light-reflective beads that cover its entire length. The cloak is also fitted with cameras that project what is at the back of the wearer onto the front, and vice versa. The effect is to make the wearer blend with his background.*[8]

The science behind this is, in some ways, quite easy to understand, although it requires a radical shift in how we utilize our laws of optics. It has to do with light refraction and the visible spectrum through which humans literally "see" the world. The portion of wavelengths in the electromagnetic radiation that humans perceive is known as the visible spectrum. The visible spectrum is actually quite a narrow band of the entire spectrum, with (human) visible wavelengths ranging from about 350 to 400 nanometers (violet and purple light), to 700 to 750 nanometers (deep red light).

As science has already demonstrated, what is beyond our narrow visual range does not exist simply because we cannot see it or utilize it without aid—or have yet to understand how to work with it. Consider that infrared wavelengths lie between the visible spectrum and the invisible microwave wavelengths of the electromagnetic spectrum. We cannot see infrared wavelengths, and yet along with other animals, plants, stars, planets, and so on, we emit these "far" infrared wavelengths (those farthest from the visible spectrum) as thermal output. "Near" infrared wavelengths, closest to the visible spectrum, are how our television remote communicates with the television—among a host of other uses. An interesting property of near-infrared light is that it has a longer wavelength than visible light (measuring approximately 750 nanometers to 1 millimeter), so it behaves differently when it encounters objects that get in its way.

Working with the larger light spectrum, scientists at UC Berkeley have been devising computer simulations that would allow them to alter the direction and properties of visible and invisible light. A recent article in the *UC Berkeley News* states:

Scientists at the University of California, Berkeley, have for the first time engineered 3-D materials that can reverse the natural direction of visible and near-infrared light, a development that could help form the basis for higher resolution optical imaging, nano-circuits for high-powered computers, and to the delight of science-fiction and fantasy buffs, cloaking devices that could render objects invisible to the human eye.[9]

What are these 3-D materials? They are *metamaterials*: substances that have optical properties not found in nature.

> *Metamaterials are created by embedding tiny implants within a substance that forces electromagnetic waves to bend in unorthodox ways. At Duke University, scientists embedded tiny electrical circuits within copper bands that are arranged in flat, concentric circles (somewhat resembling the coils of an electric oven). The result was a sophisticated mixture of ceramic, Teflon, fiber composites, and metal components.*[10]

Materials found in nature have a positive refractive index: "a measure of how much electromagnetic waves are bent when moving from one medium to another." All metamaterials have *negative refraction*. This property is derived from its structure rather than its composition. In order to achieve negative refraction, the metamaterial's structural array "must be smaller than the electromagnetic wavelength being used." It isn't surprising then that scientists have thus far had more success manipulating wavelengths in the longer microwave band.[11]

Michio Kaku, in his book *Physics of the Impossible*, explains further that "metamaterials can continuously alter and bend the path of microwaves so that they flow around a cylinder, for example, essentially making everything inside the cylinder invisible to microwaves. If the metamaterial can eliminate all reflection and shadows, then it can render an object totally invisible to that form of radiation."[12]

Scientists currently working with these metamaterials would be the first to tell you that these new engineered materials offer the possibility for us to have control over matter in seemingly magical ways, and to rewrite the laws of optics or acoustics in ways we had not even conceived of previously. This is the model that is Matrix Energetics: the rules are only suggestions, and there is so much more for us to learn, explore, understand, and imagine as possible. From our currently coveted belief structures, we code invisibility or levitation or spontaneous bone realignment as miracles and out of the realms of possibility. But even without having a complete scientific explanation at hand for these so-called miracles, we have already delved into the magic and mystery of the world and our consciousness enough to know that we are only on the tip of discovering the unknown but not the unknowable.

A Very Dark Matter Indeed

In our solar system, the movement of the planets closely conforms to Newton's gravitational laws. Because of this, it was assumed that the farther we moved out into the universe, the slower the rate of the movement in the spiral arms of the galaxies would be. In the late 1920s, astronomer Jan Oort was surprised to measure that orbital velocity of the stars in the Milky Way do not decrease in velocity the farther they get from the galactic center. In 1933 Fritz Zwicky noted the same anomaly in galaxies forming galactic clusters and suggested it was due to an unidentified "dark matter," which balanced out the mass at the centers of galaxies.

Astronomers have now calculated that based upon the predictions of the big bang theory, less than 1 percent of physical matter can be accounted for. That's right: 99 percent of the known universe cannot be accounted for. And I thought I had math problems! This invisible matter was dubbed "dark matter" by scientists not because it was evil but because it could not be measured and is not currently thought to be part of the electromagnetic spectrum.

Since dark matter and energy fill 99 percent of the universe, it would be simplistic to assume that this percentage would be composed of only one type of particle. It appears that dark matter is composed of massive superparticles. There is probably a great diversity of particles and energies included in dark matter and energy—including exotic particles and energies that escape the imagination of both physicists and metaphysicists at present.

Author Jay Alfred speculates that perhaps physicists' dark matter and metaphysicists' subtle matter and energy, sometimes called the luminiferous ether, may be one and the same. Alfred further speculates that *chi* and *prana* are likely to be categories of dark matter. Alfred wonders if pranic globules, which are more visible on sunny days, might be a form of energy exuded from the corona of an invisible sun composed of dark matter. H. P. Blavatsky, cofounder of the Theosophical Society, often spoke in her writings of a mysterious sun behind the sun. Tesla also spoke of the zero-point energy, a mysterious sun behind the sun that was the source of the zero-point field.

I believe that the bioplasmic field probably contains large amounts of what scientists call dark matter. This dark matter could just as well be called invisible matter since our current scientific instruments cannot detect it, but due to the gravity-lensing effect, we know that it must be there. I have read many books on the rather esoteric subject of invisibility in an attempt to formulate a conceptual model for how something like it might be possible. Human invisibility has been written about for centuries. Many of the sources that I read, such as *The Golden Dawn* by Israel Regardie, had a common thread: wrapping the body in a dark mist or cloud that would render the practitioner invisible to the naked eye. I believe that this dark mist that is spoken about may very well be a "cloud" of dark matter, which through meditation, visualization, and concentration can be cultivated and utilized.

This dark matter is also probably synonymous with the concept and phenomena of *chi* or *prana*. There are many ancient traditions and practices that have to do with the cultivation and storing of *chi*. I believe dark bioplasmic matter may be, over time and with practice, cultivated and stored in the auric field. This dark matter, when enveloped around the practitioner, could then be at least theoretically utilized to render them invisible.

What Is Plasma?

In his book *Between the Moon and Earth*, Alfred writes:

> *Plasma, which is rare in our immediate environment, is the dominant state of matter in the visible universe. Plasma makes up more*

than 99% of our visible universe! The visible universe is, in fact, a plasma universe with bodies of plasma in a pervasive cloud of diffused plasma.[13]

A plasma is produced when positive and negatively charged ions become separated and generate electric fields. The field accelerates the charged particles to high velocities, thus creating a dense magnetic field. It is thought by some that these fields are composed of dark matter. Many scientists believe that dark matter is found largely in some form of plasma.[14]

BIOPLASMIC FIELDS: THE HUMAN AURA

The bioplasmic body, or subtle energy, is a complex structure that penetrates and surrounds the physical body, i.e., both internally and externally. Ancient esoteric tradition speaks of an ethereal or pranic body, a concept that is very close to that of bioplasma and the bioplasmic body.[15]

Plasma metaphysics is the application of plasma and dark matter physics to the study of our high-energy subtle bodies and their corresponding environments . . . Many metaphysicists report that the ovoid of the aura is wrapped by a membrane or a sheath. Surface currents on the shell or sheath separate the magma ovoid from the surrounding magma environment. The sheath acts as a protective electromagnetic shield whose strength and polarity can be adjusted by an act of will by the owner of the body, using focused visualizations and other techniques common in meditation. *This offers protection against electromagnetic and other intrusions.*[16]

All these properties were described and documented more than two thousand years ago, mainly in Hindu and Chinese acupuncture literature, but they were also alluded to in Buddhist and Christian scriptures and literature—long before the age of electricity and magnetism, which only emerged during the eighteenth century.

ACHIEVING HUMAN INVISIBILITY

In the late nineteenth century and early twentieth century, there arose a philosophical and spiritual development organization known as the Hermetic Order of the Golden Dawn. (A modern order continues to this day.) In its original manuscripts, there is something called a "Ritual of Invisibility." The ritual offers specific, detailed instructions in order to achieve the "shroud of invisibility"—the magic of concealment—also referred to as a cloud or veil. More than two hundred years earlier, Paracelsus expressed the idea of concealment in his *Philosophia Sagax*:

> *Visible bodies may be made invisible, or covered, in the same way as night covers a man and makes him invisible; or as he would become invisible if he were put behind a wall; and as Nature may render something visible or invisible by such means, likewise a visible substance may be covered with an invisible substance, and be made invisible by art.*[17]

This idea appears frequently in folklore as well. The cloud is usually referred to esoterically as a garment of some kind which, when worn, conceals the hero from view. Of course, what comes to mind with this idea in our culture is Harry Potter's cloak of invisibility.

If electrons have the power to absorb light photons when they are bound into atoms, there is no reason why they should not have the same power when they are free. And the cloud is just a cloud of free electrons or bioplasma. Since the energy gap of the cloud is apparently quite small, all the photons that enter it are absorbed, whereas some are usually reflected from more ordinary substances. And with zero reflection, we get zero visibility.

Since electrons are the building blocks of the atom, and since the atom is the building block of matter, one can easily see how a "cloud" of electrons could be formed into solid matter by mind power. But what you may not see is how such a cloud could make a human being invisible. The fact that the cloud is a cloud of electrons is the key to its power of making things invisible. Scientists know that such a cloud of electrons will absorb all light waves entering it, reducing the magnitude of reflected light to zero and effectively concealing whatever it surrounds.[18]

CREATE THE CLOUD EXERCISE

Here is an exercise for developing invisibility, manifestation, and other *siddhis* (or powers). Author Steve Richards gives an exercise similar to what I have outlined here, but my reference for this exercise is taken from my personal experience with an alchemical exercise I have worked on for a number of years. This was given to students of my spiritual teacher and friend, Elizabeth Clare Prophet. My original exposure to the cloud exercise came from the book *Saint Germain on Alchemy* by Mark Prophet.[19] The exercise below is my own adaptation of what I learned from my spiritual training, as well as elements that Richards elucidates in his book, which I have excerpted in part for elements of the exercise below.

STEP ONE

The first step is to construct your laboratory. This laboratory is none other than your sacred space in the torus field of your heart. When you have developed a momentum at this, all you will have to do is think about moving into the alchemical laboratory of your heart and you will be there.

The cloud is composed of subtle etheric substance. According to Saint Germain, you can concentrate the cloud through a passive act of will. He states that the cloud should be formed of a milky white radiance similar to the star clusters of the Milky Way.

STEP TWO

With that taken care of, the next step is to sit quietly and comfortably, and direct your eyes to some single place in the room. This is necessary so that the cloud may collect at the place where you are staring. The effect of directed attention is cumulative. The longer you look in the same direction, the more definite the cloud you are building becomes.

STEP THREE

By slightly defocusing your eyes, you will enter into what mystic Don Juan's favorite student Carlos Castaneda calls "Second Attention." This altered state will enhance your ability to see the cloud. Collecting the cloud is of no use to you if you cannot see it after it has been collected. Therefore, this defocusing is absolutely essential to the technique. Now let your attention drop down out of your head, into your chest, and then into the space of your heart.

STEP FOUR

Relax, staying focused in a calm manner on your intent for the cloud to be brought into manifestation. In order to enhance your ability to concentrate the energy, mentally circumscribe a ring of light around you at the height of your chest. This ring will aid you in your efforts to condense the force of the cloud.

STEP FIVE

When you think you are getting results, it is time to intensify the building of the cloud. Let the cloud expand out from the torsion field of your heart in all directions simultaneously so that the force of the cloud pushes against the barrier of the ring. Begin to draw large handfuls of invisible or dark matter into your ring from the surrounding atmosphere.

You may glance up above the cloud and then bring your eyes down, willing that the energy above the cloud be added to the energy of the cloud itself. Then do the same thing by glancing below the cloud and to either side of it. Remember as you do this that your willing is not to involve eyestrain in any way. The willing must be entirely mental. Remain relaxed and above all, "stay out of your head"!

STEP SIX

Once you have formed a cloud that is quite definite and that contains a great deal of dark matter, the final step is to draw it around yourself and blot yourself out of view. Once again, the technique is just the logical result of everything I have said thus far. You must produce a cloud large enough to enshroud the human body, and then will that it come toward you and surround you. Become lost within the billowing folds of its milky radiance if you wish. The practice of forming and learning to sustain the creative energy of the cloud will greatly enhance the power of your creative manifestations. With practice, you may be able to cloak yourself within it and become invisible to the naked eye!

ACCESSING THE UNIFIED FIELD OF CONSCIOUSNESS

When we are connected in this way, we have created a special unified field of consciousness. Actually, what we have done is create a special relativity that, through the heart, accesses the unified field of consciousness. At this point, everything becomes a pattern of oneness, and if anything does not match this pattern, simply rotate the consciousness until it matches. Take all these complex thoughts, interactions, and complexities and get into the simple state where change occurs. It is what we call in Matrix Energetics "shifting your observational frame of reference." There is actually a workable physics that supports all this, which has been utilized in projects such as the Philadelphia Experiment.

In yogic tradition, the mastery of invisibility (as well as other siddhis) is said to be controlled by the power of the heart chakra. I believe that the heart chakra possesses a right, or clockwise, spin. It is my theory, after much contemplation, that the bioplasmic field of the aura possesses a counterclockwise spin. In order to get the changes we are talking about, like in the Philadelphia Experiment, you must shift the frame of your reference to your heart. *When you master the unified field of the heart, you can actually fold space-time locally.*

You can learn how to shift within yourself, into the field of your heart chakra, and from there you have access to interdimensional space. When you start to shift that torus tube, the torsion field of the heart's energy starts to spin in a clockwise direction. As the energy reverses direction, the torus starts to expand. (That is exactly what the magnetic field or Tesla coil would look like.) Now what you do is start spinning the auric field rapidly in a counterclockwise direction simultaneously. This activates a unique bidirectional set of counter-rotating bioplasmic energy fields. Now you have two points of reference for hyperdimensional space, and that is enough to collapse space-time where you are.

Add to these counter-rotating auric fields the mastery of the previous cloud exercise, and I believe you can attract dark matter from the realms of the bioluminescent ether to your spinning field. This dark matter, with visualization and practice, can enfold your human aura. Since this bioplasma is invisible, you are cloaking yourself in an invisible substance, which is increasingly magnetized to the expanding, spinning field of your heart. Perhaps when these counter-rotating energy fields reach the appropriate vibrational rate, your aura might radiate bands of energy into the ultraviolet spectrum, which is beyond the bandwidth of human visible perception.

You are now starting to understand how the power to do many of these things resides within the unified field, the torsion field of our heart. Basically, torsion fields interact with the spins of particles. In the atom, the nucleus, the protons, and the neutrons all have spins. Torsion fields create a hyperdimensional geometry, which has access to extradimensional realities. When you master the unified field of the heart, you can actually fold space-time locally. It is all based upon what you engineer in consciousness.

What you engineer in consciousness is then what you can rely upon. You engineer a reality. That is the principle behind engineering space-time and why Einstein's unified field theory—though incomplete—was engineerable. When you are holding things in a dimensional matrix within your heart, that thought-form or concept can imprint upon the counter-rotating fields formed by the torus field of your heart. This electromagnetic spherical field of the heart can then interact with the bioplasmic fields of your aura. This consciously engineered interdimensional matrix can access hyperspace dimensions. This is Philadelphia Experiment technology.

TALES OF INVISIBILITY FROM THE FIELD

I was on my way to the Children's Program seminar in Los Angeles, driving 80 miles an hour in a convertible, holding on to my Indiana Jones fedora so it would not blow away. Until then I had been writing this book nonstop, holed up in a beautiful resort hotel in Redondo Beach. In order to attend the program, I had to drive an hour each way, and I somewhat resented the intrusion into my writing time. My guides had told me that it was important for me to attend, however, so I was on my way. "What could possibly show up at a Children's Program that I would need for the book?" I wondered. When I arrived, my daughter Justice was already presiding over a group of youngsters and their childlike parents.

A frequent seminar attendee named Alejandro approached me. He had brought his whole family to this seminar. He introduced me to his family and said he was delighted to see me. He then said mysteriously, "I have something that I must tell you. It makes no sense to me, but my inner guidance insists that you must have this information."

He went on to tell me an incredible story. A number of years ago, Alejandro lived in a house that could only be accessed by crossing a suspension bridge. His cousin, who was staying with Alejandro, was in trouble with a gang that wanted to kill him. The gang came looking for him at Alejandro's house. Hearing the gang's approach, the cousin ran out of the house and started to cross the suspension bridge just as the gang came around the house looking for him.

The cousin knew with certainty that if he were to get caught by the gang on the bridge, he would have no escape and would be killed. It was a really long suspension bridge, so he reached into his pocket and pulled out a prayer to Mother Mary (the *Magnificat*). As he prayed to her, a bubble suddenly popped up around him and this strange green fog filled the space in the bubble, and he became invisible.

The gang members looked at the suspension bridge where they had seen him running, but they didn't see anybody there. The cousin got across the bridge and popped back into visibility again. They saw him then, but he was too far away for them to catch him. They were baffled as to how he could have disappeared and then reappeared. They never bothered him again. I heard this story and thought, "Oh, I am going to have to include that in the book—that's why I'm here. Here is a personal example of the Philadelphia Experiment."

WHAT? AM I INVISIBLE OR SOMETHING?

My son Nate and I have programmed our Matrix Energetics medallions for invisibility. We did this last year before a seminar with more than five hundred people coming into town to attend. I was overwhelmed and a little out of sorts because all these people would be focusing on me. I felt funny, so I created a field around the medallion that enabled me to step into a place where I did not feel the energy. I could rotate my personal energy field out of phase with it.

Nate and I went to our favorite Mexican restaurant, where they know us and know what we are ordering before we are seated because we eat there so frequently. We sat there for forty-five minutes before we realized why nobody had come over to take our order or acknowledge our presence. We had programmed our medallions for invisibility. So we mentally clicked off the program to allow us to be visible, and almost instantly a server came to our table and said, "Hello, Dr. Bartlett. When did you arrive?" It can be weird—and useful.

A SEMINAR PARTICIPANT'S STORY ON INVISIBILITY

During my first conference, I thought it wise to share some of what I found interesting prior to the ME training with others. I had a picture, which I had downloaded off the internet, of a young man in Japan, who was wearing a light green parka, facing the observer or camera, while standing on a street. The parka was coated with a film referred to as "the Invisibility Cloak." This film is designed to allow light to pass through it, and that of any surface to which it is applied. In the photo, you could see three other young men walking at a distance behind the person wearing the parka. Their images showed clearly through the parka, as did the street and all the associated physical details.

I gave a copy of this to Richard, showed it to several others, and gave another copy to a lady named Christa. Then the strangest thing happened. She told me she had had an experience where a women raised her hand in front of her, and Christa could see through this woman's hand. To be certain that this was a real event, she asked the woman to raise her hand a second time. And again she was able to see through the person's hand as if it were not there.

When Christa told me this story, the three of us were about to eat lunch. I always bless water as part of my eating ceremony, and as I was doing this with my eyes closed, I noticed that I could see the water bottles and glasses clearly with my eyes closed. I could also see through the space these objects occupied and out beyond this construct to another indeterminate zone. "Houston, the Matrix has landed, and it is trying to grow roots!"

—*EP, Matrix Energetics Certified Practitioner*

HIDING IN PLAIN SIGHT

A female seminar participant once came up to me and told me she was so glad that I was teaching about invisibility. She had previously been to a seminar where I spoke about the science of the Philadelphia Experiment and its possible application as a spiritual technology. I had given an exercise about accessing the torsion field of the heart, similar to the one included in this book. I then discussed how counter-rotating torsion fields in the human aura could conceivably duplicate the technology of invisibility demonstrated in the story of the Philadelphia Experiment.

She went on to tell me her story. Sometime after the torsion-field seminar, she was walking down the street in her hometown and saw her ex-boyfriend heading toward her from the opposite direction. If ever there was someone she wished to become invisible to, it was him. Her breakup with him had not gone well, and he carried a grudge. Remembering what I had taught her, she dropped down into the sacred space of her heart and began the exercise of creating counter-spinning twin torsion fields. To her absolute amazement, her ex-boyfriend walked right by her and never even noticed her! She was so grateful for the information that I had shared with her about invisibility.

CLOSING THOUGHTS ABOUT MIRACLES

I am not saying that you are going to be able to become invisible just by reading the information in this chapter or perform miracles of healing just because you have a desire to do so. It does not matter. What matters is if one person can do any of it, and can do it one time, then you have access to it, and it becomes more and more likely as a weighted probability that you will be able to do it as this field continues to expand and grow.

There is a true religion and true science, but they are one and the same: the unified field of consciousness. Now, understanding that, you have broken the paradigm. You are now ready to actually play because from this point of reference, it does not matter what you do. You are making it up. If you are making it up, all you have to do is be congruent and coherent.

This is spiritual technology that then is engineerable to be actually workable as a physical manifestation. Tesla and others have done it. By the way, when I say "spiritual technology," there is nothing *but* spirit. There is no physical. There is no matter. It is just spiritual and consciousness through Heart. That is it.

17

SOLVING THE MYSTERY OF LEVITATION

Matrix Energetics seems to mine the same scientific terrain as that of the secret aether physics. If our secret government has spaceships that can rise up in the air, using secret or black-op physics principles, these elite technologies do not seem to rely on the crippling assumptions promulgated by Einstein's special and general theories of relativity. After reading all that I recently have on these subjects, I suspect relativity is a well-sustained scientific blind alley. I am thinking there may be principles involved that are discoverable and engineerable, as well as reproducible. If my research is any indication, we have been secretly employing these ideas and technologies at least since the early 1940s.

Einstein completed a version of his unified field theory, which united electromagnetism and gravity, in 1928, and he presented it in Prague. What he did to unite electromagnetism and gravity was one simple little thing: he took an idea of a fifth-dimensional geometric space by Theodor Kaluza. Kaluza had told Einstein that if you take length, width, and depth, and add in the fourth geometry of space and add in time, you actually get a unification of electromagnetism and gravity.

What Swedish physicist Oskar Klein computed was that this fifth-dimensional space was so tiny that it was at Planck's length (10^{-31}). This fifth dimension was computed to be so small that it existed on the point of Planck's constant outside the dimensions of space-time. It folded around every point in space-time. It is believed that is where electromagnetism is, outside space-time, and that it comes in as an activated potential.

The only glitch in Einstein's theory was that he could not get the strong and weak nuclear force to fit the model, so he eventually withdrew it from public view. It was this incomplete yet engineerable unified field theory that may have become the scientific underpinning for the Philadelphia Experiment.

Tom Bearden thinks that what we call "gravity" has its origin in the vacuum, where it is a strong force. When you flip the forces of electromagnetism, you get gravity and its opposite: antigravity! When you tap into the electromagnetic flux within the vacuum, you have the potential to create enormous electrogravitational waves. These waves, which Tesla also discovered, could be used to artificially engineer and create a workable source of antigravity to power UFOs: man-made, intelligently controlled, highly advanced flying conveyances. You also get the ability to engineer specific electromagnetic waves that can be sent instantly through the vacuum to great distances, targeting a distant location.

We have never been able to figure out what gravity is. We can't find it. We theorize about it, but we can't unite it with electromagnetism. What if there is only the one energy and everything else is just a subset of that one force?

Basically, gravity does not exist as a separate force. If it did exist, you'd think we would have found it by now. But we can't find it, and we can't get it in any of our mathematical formulas. The only thing we can get is a curve in space-time with a very large body.

Conventional physicists say you cannot curve local space-time. Perhaps we cannot do it because we are using the wrong mathematical models. Maxwell figured this out—and then his model was imprecisely replicated by Heaviside and it got lost. When you embrace any model based on limitation, your perceptual bias becomes "it cannot be done." That does not mean it cannot be done. It just means it is not likely based upon what you have as an equation for your reality.

DIMENSIONS OF REALITY

Review the qualities I have listed in the chapter concerning torsion fields and their properties if you doubt this. Here is what I think: electromagnetism is actually an effect, and, in fact, there are no forces. There are merely potentialities that come from the zero-point field. Therefore, electromagnetism and gravity are one and the same thing, which is why you can't find it. But gravity resides within the zero-point field. Electromagnetism is the force that we see here. You can unite gravity and electromagnetism through that fifth-dimensional space, which wraps around every point in fourth-dimensional reality. Electromagnetism actually comes from outside of our fourth-dimensional reality, and it is an effect that we see.

If everything you see in your personal world is "effect" and you can get down to "cause"—which is the unified field of the heart—then you are one with all those forces. You unify with those forces. What happens if you unify with those forces? Well, first consider that you are dealing with polarities in space-time. What if you flip gravity in the unified field? What do you get? You have antigravity. I think the saints could do these things and that they were accessing these extradimensional states through the force of love. Through love they accessed the extradimensional state that is held within the tube torus in the field of the heart.

Jesus walked on water, demonstrating mastery over gravity. From the point of the torus field of the heart, gravity, or antigravity, becomes just a matter of flipping the charge. Thought is what causes the unification of these forces. Therefore, whatever you think and create a polarity for is what you manifest. The key is to get your thoughts and feelings subjugated to the power of the field of your heart. The magnetic field of your heart is actually stronger than the electrical field generated by your brain. The field of your heart is a true torsion field. From the point of the heart, when you truly make your abode there, you can potentially master all of time and space.

This technology of consciousness resides right within the torsion field of our heart linked to our bioplasmic energy fields. I believe it is the physics Jesus utilized. It is the physics of Nikola Tesla as well. So that is what we are beginning to teach. The cool thing about this is that not too many people are teaching this. So what do we have? We have an unpolluted morphic field that we can build however we choose.

I am not saying we should all try to levitate. It may have been a good idea when I fell off the stage. It would have been a pretty useful miracle. Workability is the key. But if we are willing to entertain these ideas, then what happens is we create a dynamic in the morphic field (called Matrix Energetics). If we do this, miracles are more likely to show up because we know there is a science and logic and there are procedures and it has all been done before. If it has been done before, it can be done again. It does not matter if it should be done or if you need to do it. It is a principle that exists that is definable, observable, and reproducible.

I am confident that certain esoteric spiritual principles and practices are applicable to the phenomena of levitation and invisibility and the performing of miracles. If the black-ops physics allows for incredible secret technologies, I believe that the physics behind these spiritual and material manifestations are one and the same. I submit to you the idea that if one person can float in the air, all people can float in the air. I am not saying that we are going to be having seminar participants floating around in the air anytime soon—just that it's possible.

Have you seen the Buddhist meditation posture where you close the loop of your conscious attention? This process is represented and perhaps engineered by the union of your thumb, index, and middle fingers joined together. What if this *mudra* does more than just direct your focus inward? What if it literally directs external EM energy inward? Now what is all that external energy? Well, it is not gravity—which is a very weak force of a magnitude of 10^{-42}.

However, what is the mirror of gravity? That would be electromagnetism, which is 10^{42}, coincidentally enough. Do you see a yin/yang sort of thing occurring here? Now, if you invert those equations, then you get gravity at 10^{42} in the vacuum and 10^{-42} here. If you do that, what do you do? You would float up, or levitate, into the air.

In Hindu esoteric anatomy there are two channels or forces called the *Ida* and the *Pingala* in our nervous systems, which run up both sides of our spines. What if those energies are your forward wave and your time-reversed wave? When you focus on these energy channels and you combine them in the central spinal channel, or *shushuma*, perhaps you get a scalar electromagnetic wave.

In other words, you get two waves at 90 degrees out of phase with each other that yield a sum-vector of zero. In other words: they are not detectable by our current technology but contain the force of the vacuum. That would be why raising Kundalini can generate powerful effects, including levitation or walking on water, healing the sick, and perhaps even the ability to raise the dead!

How is all this weird science useful for you? Anything that can be used as a weapon can also be used to heal. Time-reversed energy has positive effects such as healing, stopping decay, increased longevity and vitality, the ability to levitate—and, oh yeah, miracles! So then what are you doing? You are tapping the virtual energy of the vacuum. If you can harness this energy from the vacuum, you have produced a source of free energy. You can engineer the vacuum's energies, structuring the virtual photons present into an artificially engineered template of action. You can simply flip the polarity or phase of a disease pattern, and a healing miracle can be the result!

Yogananda said in *Autobiography of a Yogi* that the advanced yogi could access enough light energy to run the city of Chicago. This energy is in the zero point. You can't get to the zero-point field unless you go through the higher computational capacity of the right brain. And the right brain has direct access through the field of the heart.

These spiritual and energetic technologies have been around for maybe two hundred thousand years, and they possess a really big morphic field that almost no one is tapping into. When you deliberately tie into the morphic field of this kind of ancient power for the purpose of healing the sick and helping the planet, you are using this spiritual technology of consciousness in a way that supports the best qualities in mankind. You are becoming a true Light Worker and world server.

DR. HECTOR GARCIA'S LEVITATION STORY

When I was growing up, my mom met my stepfather. He was a professor at Cal State Los Angeles and an American gentleman. One afternoon, he took me to Pasadena to the Self-Realization Center. I thought this was pretty cool. I asked him what the center was about. He said it was about prayer and meditation and devotion. He said to relax and feel good. He went into a room with some adults and told me to go sit in the garden. I thought, "OK, that is cool. I can do that."

I had always been interested in yoga and meditation. I had read several books and remember that the first one I read was on Hatha Yoga. I would see the yogi sitting there in lotus position, and I remember thinking that I wanted to do that. No one told me that I could not. So I picked up the book and then started doing it.

So when my stepdad and I went to the center, I was just doing my "thing" out in the garden. I recall vividly that it was a sunny Sunday afternoon, and I was just sitting in my lotus position, meditating. When I opened my eyes, there were people all around me, looking at me. Naturally I felt a bit funny and wondered, "What are they looking at? What is so strange about what I am doing? I thought you are supposed to sit in lotus position, raised off the ground a little bit."

Someone asked me, "Who taught you this?" Innocently I replied that I thought you were supposed to do this, and that I had read it in a book. I was told no, and then I fell down.

Years later when I was in high school, I ran cross-country. The night before a race, I would go into a trance and picture the race the next day. I would picture a Kriya Yoga star with Christ in the middle. I would meditate with my back to the wall for about twenty minutes. Invariably, my sister would intervene and tell me I was strange because although I would start with my back against the wall, at some point my back would be turned and I would be in lotus position, facing the other way. I was off the ground and turned around.

I still remember being able to levitate, but I cannot do it now.

This experience Garcia shares breaks your paradigm. If average people can levitate, I suspect that momentum is the reason. It is just about dropping into the state and doing what has already been done before. If you are supposed to be able to read a book and do it, then that is why we are writing that second book. There will be people who read *this* book and they will do what we talk about. Then those people will link up to a grid or morphic field where more and more people will be able to do it, and pretty soon the world can shift.

Have you heard stories about Tibetan monks being able to levitate and do what in general seems to subvert the known laws of our current physics model? If they are indeed doing these things, then our physics model is incomplete at best, and possibly even deliberately deceptive or wrong! I am not saying that these siddhis should be a goal of Matrix Energetics, and I don't need to go to extreme examples in order to understand the principle. Do you understand this? It is practical to do this. Unless you really break your habitual patterns of thinking, your coefficient bonds with normal consciousness, then what happens is that you do not exceed the orbit of your limited awareness.

Review the qualities I have listed in this chapter concerning torsion fields and their properties if you doubt this. Here is what I think: Electromagnetism is actually an effect, and in fact there are no forces. There are merely potentialities that come from the zero-point field. You are now starting to understand how the power to do many of these things resides within the unified field, the torsion field of our heart.

Miracles have occurred before and can occur again. It does not matter if one should occur or if you need to "do" a miracle. Again, it is a principle that is definable, observable, and reproducible. Do you get this? Remember that you can step into the dynamic field of consciousness that Matrix Energetics has built because we have said, "What if there were no rules?"

One of my students told me about some friends who were visiting India. The friends became so joyful at one point that they began to float off the ground. This was not a trick. It was real: Again, it is a principle. I saw the picture of this. My point is very simply that you do not need to be able to levitate or walk on water or even perform miracles. Miracles will happen when you stop trying to make something happen. You are the miracle.

18

ARCHETYPES: MAKING A LOVE CONNECTION WITH THE STUFF IN YOUR HEAD

As you begin to work with some of these inner processes, pay close attention to the types of energies and characters that bubble up from your subconscious. Keep track of the unique symbolism that your mind employs to communicate with your conscious awareness. By doing this you will begin to strengthen the connection with your greater self-awareness.

Consider keeping a written diary of the images that you encounter repeatedly and try to figure out what is being represented by them. A lot of guys won't ask for directions if they get lost; don't be like that with your own mental processes. If you get lost or do not know what something means, just ask! That is so simple that at first you might not think that it will work. Trust and ask something simple such as "What does this mean?"

ARCHETYPES AND THEIR USEFULNESS AS A HEALING TOOL

My friend Dr. Hector Garcia uses his intuitive gifts to detect energetic imbalances in the body—from the cellular to the quantum level—with amazing accuracy. He talks about archetypes and their usefulness as a healing tool:

When I teach about archetypal patterns, I like to do them with physical structures. For instance, if I am working with a client with scoliosis (an abnormal or exaggerated curvature of the spine), I sometimes visualize or see a ruler lined up against their spine. Often instantly, the act of superimposing an image of this nature will convey to the energy field the desired correction, and the scoliosis will just be gone. Sometimes you have to be careful when working with people with things like scoliosis because they can occasionally become upset when their condition changes so radically and quickly. You see, they are not used to being "straight" and the sudden change can occasionally be disorienting and upsetting to them.

I had one client who came to me, and her brother was a chiropractor who had been treating her for her scoliosis. She was told that I could help her. I did the corrections that showed up in her case, and afterward her spine was no longer twisted. She came back a week later, mad at me because her clothes no longer fit. Go figure!

SUPERHEROES AS MYTHOLOGICAL ARCHETYPES:
GREEN LANTERN AND THE AWESOME POWER OF INTENT

One useful thing that I have always done, being a recovering geek and comic book fan, is populate my subconscious terrain with noble figures such as Superman, Batman, and Spiderman.

Green Lantern is the coolest superhero ever because the greatest force in the known universe powers his ring: *the power of intent!* This pure intent, when shaped by a particular thought, takes on the form and activity of the ring bearer's directed focus. Whatever you think of in enough detail will manifest and can be wielded by Green Lantern or even yourself! Green Lantern supports and protects the forces of good and should not be confused with that other ring bearer, Sauron from J. R. R. Tolkien's epic, *The Lord of the Rings*. The only thing Green Lantern's ring won't work against is the color yellow. I have always taken this to represent how our own fears or cowardice can stop our intent from manifesting. As the Ascended Master Saint Germain has stated on such subjects, "Fear is the enemy of the Alchemical experiment!"[1]

At one point recently in the last year, I decided to mentally make and wear a copy of Green Lantern's power ring. Now, to many of you, this may seem silly, but instead of playing with tanks and guns when I was little, I played with superheroes. My mom even made me a complete Batman costume.

At a recent seminar, someone gave me a sizable sum of money so I could buy a very cool Batman outfit from eBay that was made to exacting detail. I was only partly joking when I told seminar participants that I had a unique weight loss plan. I was going to buy a Batman suit, complete with rippling torso and leg muscles, and hang it in my closet. I now own an exact copy of the Michael Keaton *Batman* movie costume and recently wore it onstage at one of my seminars.

SUPER(MAN) POSITION IN REALITY

There is a cost to emulating your heroes too closely. When I was a little boy, I was really fascinated with George Reeves's portrayal of Superman. (Of course, at my young age I just thought that he really was Superman!) I was so taken with Superman that I drew a red "S" on my T-shirt and my mother sewed a red cape for me. I think she was a little worried that I was going to try to jump off the roof of our garage, trying to fly. At one point my mother presented me with some horn-rimmed eyeglass frames. She quite logically pointed out to me that if you are going to be Superman, then you have to be Clark Kent as well.

It was a good argument, and so I started wearing the glasses when I wasn't running about in the identity of the big "S." There was only one little problem with this good-natured and entirely reasonable suggestion. My dad wore glasses, and I looked somewhat like a younger carbon copy of him. I think I might have inadvertently given my subconscious the wrong commands because within a matter of weeks my vision deteriorated and I actually needed glasses to see clearly. So be careful what you agree to and how it can limit your perceptions and abilities in life. I still wear glasses or contacts when I am teaching onstage, but you could say that *I see things differently these days.*

In truth, I know a number of people who can see things beyond the spectrum of normal conscious awareness on a daily basis, and they do just fine with that reality. As long as you don't think that because you can see something you are always responsible for doing something about it, you can function in that manner. Just because the world is tragically teeming with sick people doesn't mean you can heal all of them. You take each day as it comes and live in the Grace of the moment. My friend Mark sometimes sees diseases and conditions in people. Simply because he sees such things doesn't necessarily suggest that he can do something about what he sees—at least not in every case.

I will never forget how one day in the office, near the lunch hour, Mark was going out the door and I was about to start an appointment with a new client. As Mark exited the building, he handed me a folded note. I unfolded the paper. On it Mark had written the words "Be sure to question her about the family cancer pattern." Sure enough, when I asked the client about this very issue, she provided me with an in-depth picture of how cancer was indeed a factor in her family genetics.

This is not information she had listed on an intake form. Mark had "seen" this by noticing energy patterns in her auric field. But he misses things sometimes, which is why we *always double-check everything we can through conventional medical diagnostics*, which, as doctors, we have the ability to employ where warranted. Trust, but use every method at your disposal to verify.

One more word on this subject will drive home the point. One of my patients, who became a friend as well, told me how he had studied with a famous psychic surgeon in the Philippines. Now some of those individuals may not be for real, but my friend Arnold swears that he witnessed this "surgeon" open up a guy's chest and lay his heart right on top of it, out in the open. Time appeared to stop, and the psychic healer stated that he was cleaning out the coronary arteries and had to stop the local flow of time and the man's heart in order to accomplish this feat.

Arnold was a very down-to-earth philanthropist who was a self-made millionaire. I tended to believe the earnestness of his account, however mind-bending the details. Arnold further confided with me that this same healer said he was going to open Arnold's "Third Eye," or "Mystical Eye."

Arnold swore that after this procedure he could not go into a fast-food restaurant without "seeing" the various internal diseases the patrons were nourishing with the devitalized fare the eatery was serving. He told me that it took a long time, through disciplined meditation, to shut down or veil that ability so he could function in this realm. It has been said that without vision the people perish, and for thirty years this statement has driven my desire to be clairvoyant.[2] However, I am not sure that I want to see everything all the time, so I am still in negotiation with my sub-conscious mind concerning this issue.

SEEING IS BECOMING

At one point in his training with me, Mark entered into a very deep trance state and his "Third Eye" popped open. In a flash, he was suddenly seeing spirits in some other realm talking to him and walking toward him. I asked him if he wanted to turn off that experience, since it was a somewhat fright-ening turn of events. He wisely replied from his deep trance, "No, thank you. Leave it on, but perhaps you could install some kind of dimmer switch."

Brilliant and brave, I thought to myself as I complied with his request by supplying the appropriate suggestions. This was another turning point in his training—and he can still see spirits. The last time Mark taught with me was in San Francisco. I remember him commenting on the num-ber of ghosts he saw walking the streets. After all, San Francisco is famous as the home of the Grateful Dead!

I don't want you to think, even for a moment, that in order to be able to do things with Matrix Energetics you need to be clairvoyant or psychic in any way. This is not the case at all. It is okay to open your awareness to new abilities and experiences. Some of the things I have been talking about in the last few pages may begin to happen in your world even as you are reading this book. A natural unfolding of abilities will occur as you begin to play in the morphic field of possibility that Matrix Energetics represents.

I have heard many stories from my students and people who have read my first book without attending a live seminar. It seems that by picking up this book and reading it, you are in some instances signifying to your subconscious mind that you are ready to experience some of the things I have written about. If such things do not happen to you and you wish that they would, go easy on yourself and don't be in a hurry. Trust in the process of learning that seems natural to your pace and allow yourself to open slowly, like a beautiful flower opening its petals to greet the new morning sunshine.

When I "play" with someone, I say "this" and "that" as well. I do not even know what "that" is. Well, I kind of know, but here is the thing: if you slow it down in order to figure out what it is, then you are taking too long. If you take too long, nothing will happen because you are observing whether it is happening or not. You can become educated in not knowing. You can work with primary, heart-centered knowing coupled with secondary, right-brain, nonspatial awareness.

Jesus said, "Ask, and it shall be given you; seek, and you shall find; knock, and it shall be opened to you."[3] These and other such words from Jesus are equations in higher consciousness: that is, the Quantum Jesus again. The act of asking open-ended questions opens the doors of perception to the *All That Is*. From this point of trust you can make an assessment, not a judgment.

As soon as you judge, you collapse the wave function, and the wave function represents all possibility outside the domain of time and space.

19

PRACTICING THE ART OF INNOCENT PERCEPTION

Welcome and embrace what shows up and you will learn to see. When you look through the eyes of your opening heart, you get a feeling for what shows up. This feeling state creates the arc of contact between your heart and your desired outcome. This is part of the science of manifestation. It is a feeling state. As you begin to engage in the childlike art of spontaneous perception, be willing to suspend your rules and go with the flow.

If you do this, your brain can begin to create unique neurochemical states that can deeply alter your conscious experiences. By habitually practicing the feeling of this altered state, you can learn to create new possibilities in your life. This state of being is not unlike the process of undertaking a shamanic journey. Trust that whatever you experience is unique to you and a gift from your unconscious or High Self. Appreciate it, enjoy it, and learn from it.

I believe in everything until it's disproved.
So I believe in fairies, the myths, dragons.
It all exists, even if it's in your mind.
Who's to say that dreams and nightmares
aren't as real as the here and now?
John Lennon

THE CONTROL GROUP FOR ALTERED STATES

A gentleman was participating in one of the Matrix Energetics seminar exercises, and a large control panel showed up in the air in front of him. He did not ask for it. It merely showed up. At the time, he did not know what it was for, so he merely acknowledged it and moved on. Now, he told me, he regularly sees and converses with spirits of what some people call "the spirits of the dead." Perhaps he has had this ability all along. He came into life with the control panel set for medium and now he has moved into high gear. So trust what shows up. It may just change your life.

Tolkien's *The Hobbit* is subtitled *There and Back Again*. Some forms of schizophrenia and madness could be subtitled "there and did not make it back again." In the former, you have Gandalf the Wizard to help ensure your safe arrival. In the case of schizophrenia, you may not know you have help. Trusting in a higher power and then demanding proof can make a world of difference. God or the devil will never make you do anything against your free will. But you can create a god or devil that can enslave you to a set of choices not conducive to freedom or personal transformation. Choose wisely!

OUR PERCEPTIONS CONFORM TO OUR
UNCONSCIOUS RULES AND EXPECTATIONS

Unconscious behavior is mapped by what we call *physical laws*. These laws, or rule sets, describe what is allowed and what is forbidden. What we conceive or perceive depends on our model for how reality should be in our personal worldview. Our unconscious rule-filters determine what shows up in our world and equally what is denied conscious access. For example, rarely will you hear of a Christian who has a vision of Buddha or a Buddhist who in meditation receives Jesus Christ as his or her personal savior.

Our reality recipe is flavored by the ingredients we allow and the individual tastes we prefer. This mixture is formed from equal parts of what we know, what we knew, and what society expects us to believe. One possible definition of conform is to "co-form" or "cocreate." Collectively, at an unconscious level, we create and sustain the cultural hologram of our sensory-based experiences.

It is the unconscious box containing your assumptions, beliefs, and experiences that is the driving creative force behind each and every one of your manifestations. If you are reading this book and some of things I say seem to go right over your head, or perhaps beneath your feet, then that is okay. When you don't immediately understand a concept or idea, this can be a signal that *you are learning new information.* All our concepts are largely based upon duality and linearity: *the gospel of opposing forces.* Remember, though, the only time you want to go up against a posing force is when it potentially involves super-models!

Angels and gods and comic-book heroes qualify as excellent super-models. If you model stellar qualities in the collective and unconscious archetypes of power and possibility, you can, to varying degrees, learn to harness the naked energy of those forces. Frankly, the raw psychic energy stored within these cultural archetypes can be used in magical ways. If you don't believe me, read the first chapter of my first book and then come back to this one. Jesus, Superman, and Spiderman are all wonderful archetypes you can tap into, but the boogeyman isn't.

If you give up the sense of struggle, there is no need to enter into battle with yourself or some part of the collective psyche. Energy not expended in internal warring can be used to make more love, sex, and money—the three psychic food groups! Give up that struggle with your ego. Change the templates and patterns that are predicated on the old ways of being. You can move into new awareness patterns that provide for expanded opportunities in every aspect of your life. Expanded and altered awareness is the key that opens the floodgates of opportunity.

Here is an example of the use of archetypes from a Matrix Energetics student and "whizard":

The MD I'd helped with tennis elbow and, more recently, sinus/allergies just shakes his head and says, "Bob, that's just not right" when his problem goes away. Still not entirely convinced, he stopped by the other day, saying, "Bob, I've been getting these cluster headaches. I don't have them now, but can you prevent them?" I told him we could surely try, and I did a triangle archetype with one hand on his shoulder and the triangle about twelve inches from his head. I asked, "What would it be like NOT to have any headaches ever again?"

He just shook his head again and said, "Bob, initially I thought this was BS, but I felt something that time . . . there IS something to this!"

"Yup," I said, "Been trying to tell you!"

—*BB*

20

HUMAN INTENTION AND DIVINE INTERVENTION

For a number of years, I was involved with a mystical sect and was taught how the power of the *Spoken Word* can be used to create change.[1] If you have ever wished you could take something back right after you said it, you have a real-life example of how words can facilitate change. *Words are cups of light that contain holographic patterns and pictures.* When these auditory and conceptual templates are unlocked and set in motion, they can precipitate the things that you direct your focused intent upon.

I have noticed that the bigger and more powerful the belief system or construct, the more likely a noticeable outcome can occur in the so-called real world. One of the key concepts in the mystical sect that I belonged to is that Angels and Masters are indeed very much real and that Divine Intervention can be expected. *Expect miracles and plug into the grid space in the Matrix where miracles are the rule, not the exception.*

If you are on a bridge with your life hanging in the balance as I was, the Angel Grid is preferred to the Eeyore Grid! In other words, in times of need don't be a jackass when your life depends on your ability to reach out to something greater than yourself!

THE LABORS OF HERCULES

So you may better understand what I am saying, let me share a few examples from an earlier time in my life. In my spiritual community, one of the Masters whose intercession we invoked was the Greek god Hercules. Now, as a side note, I have always found it extremely interesting that Steve Reeves played Hercules in the movies. The original 1950s television series *Superman* starred George Reeves, and in the *Superman* movies the hero was, of course, played by the unforgettable Christopher Reeve!

It makes me wonder if the Reeves name is in some form of harmonic resonance with the Superman/Hercules hologram. I wondered if the remake of *Superman* was destined to star Keanu Reeves. But that was not to be; instead, *that* Reeves became forever identified with the concept of the Matrix itself.

What follows are two interactions that members of my spiritual community had with the morphic field of Hercules. In the first example, several community members were stranded on the road with a flat tire. Drawing upon their spiritual training, they decided to invoke the energy of Hercules to put fresh air into the tire, even though they probably had a spare in the trunk. Surprisingly to them, but not to my readers, they chanted an intense oral repetitive mantra ("thou shalt decree a thing, and it shall be established unto thee"[2]) with absolutely no discernable results. The tire remained flat!

In another strikingly similar instance on another day and a different highway, a small group of this community also found themselves stranded along the road with a flat tire. Just as in the previous story, they invoked to the spirit of Hercules for help, but this time with their *flat tire reality simulation*. In other words, although they were praying to the "gods," they were acting as if a different outcome depended on them. Within moments a bright blue Volkswagen bug screeched to a halt by the side of the road. The spiritual seekers gasped with surprise when an enormous black man got out of the vehicle, towering above them.

Without a word, the man walked to the rear of the disabled vehicle and, taking the bumper firmly in enormous hands, lifted the car several feet off the ground. One of the men in the group, immediately grasping the significance of the gentle giant's act, quickly changed the tire. With the deed done, the benevolent behemoth gently placed the car on the ground, nodded to the group, got back in his tiny vehicle, and without a backward glance sped away!

In the first example, the individuals in the story *tried to use spiritual science to trump real-world physics* without a noticeable outcome. They tried to do something that would have been easier and more practical to just do for themselves. In the second example, *there were no preconditions placed upon how help might show up—just an overriding trust that it would*.

To sum up the lesson here in another way, the indomitable Saint Germain once said to a group of devotees, "Pray as if everything depends on us, and act as if everything depends on you." Good advice by which to live your life!

Two more examples from my personal interactions with these energies should bring this point home. Many years ago, my first wife and I along with our three young children drove across the country in our old orange Suburban to attend a spiritual conference. Rain had been falling steadily in California for several days, and the roads where the conference was being held were very wet and muddy. As often occurs, nature can provide interesting tests of faith and endurance.

My Suburban became hopelessly mired in thick mud, the back wheels submerged almost two-thirds of the way up. I tried several times to get out, uselessly spinning my wheels, with increasing frustration. Resolved that nothing on earth I could do was going to make the slightest difference, I said a quick prayer to Hercules, reasoning that his legendary strength was much needed here. They say the punishment should fit the crime; I say that the prayer should fit the situation when the powers of the heaven matrix are invoked.

Then I calmly walked around the back of my truck and put my back against the rear bumper, my palms lifting up on its chrome edge. After another heartfelt entreaty to Hercules, I lifted up the car, freeing it completely from the mud. Pleasantly surprised, I was able to easily push it up the rest of the way on a thirty-degree hill. I paused to look back at how far

I'd gone, only after the Suburban's rear wheels were safely again on dry road. Again, things like this are possible but only likely if you fully occupy the belief state terrain in the Matrix where heavenly help is a reasonably expected outcome.

ANGELS MOVE A MOUNTAIN OF DIVERSITY

One final story, lest you think things like what I am talking about only happen once in a lifetime, if ever. Several years after this event, I was on staff as a doctor serving people in my chosen spiritual community. My spiritual teacher lived at the top of a winding mountain road with a very steep upward incline. One winter night I received a call from her to please come up and treat her. This was in the middle of a brutal Montana cold snap complete with blizzard conditions. The always-treacherous drive to her mountain retreat was all but impassable on this night. The road was covered with a thick, glasslike sheet of slippery ice.

As I was carefully making my way around one of the hairpin curves in my incomparable orange Suburban, I came upon a semitruck jackknifed across the road. Getting out of my vehicle, I was pleasantly surprised to see one of the community's regular truck drivers, David. He told me that he had supplies in the cargo hold of his truck that needed to be delivered to the top of the mountain that evening.

Resolved to do whatever I could to assist, I carefully inched my car around the stranded truck. Positioning my rear bumper in proximity to the front of the massive semi, David and I wrapped a thick chain around the bumpers of both vehicles. Getting in my car, I engaged the Suburban into drive, placing my foot gently on the accelerator—and promptly went nowhere!

Remembering that sometimes we entertain Angels unawares, I realized the corollary to this statement. Whereas Angels are liable to be entertained when we stubbornly try to do things without their help, it makes a lot more sense to *ask for their assistance*. Performing this action makes it far more likely in spiritual law that you will become aware of their presence and they can then help. When the student is ready, the Masters may still wait a bit in order to enjoy a good laugh, but then they will invariably appear!

Mindful of this bit of wisdom, as I called out in my mind and stretched out my feelings, I felt a connection with a benign otherworldly presence. Feeling better about the odds, I again gently placed my foot on the accelerator, easing forward. A vision of a team of bright blue Angels lifting the chain and heaving in unison flashed unbidden to my mind's eye. The chain reached the end of its tether and caught for a moment as my rear wheels dug into the icy surface of the road, seeking a firm purchase.

In the next moment, it was as if the semi and my Suburban were weightless: *the ponderous laws of physics replaced by angelic rule.* Both vehicles shot quickly and safely up the mountain pass as if launched from a giant slingshot. When I told this story to my teacher upon safely arriving at her door, she laughed heartily and proclaimed, "You are really one of the team now, aren't you?"

Getting in Sync with a Reality Greater than You

To me, being one of the team means that you are in sync with the reality you find yourself in. If you want to be rich, then you study the mind-set of people who have the beliefs, habits, and actions that have been proven to generate those kinds of outcomes. You don't sit on the street corner bemoaning your current lot in life—a fat lot of good that will do. The universe doesn't listen to the whining and complaining of your inner brat. If your significant other wants you to take out the garbage, it is easier to get you to do it if he or she asks in a pleasant tone of voice. Strangely enough, *if the voices in your head are of the pleasant variety, it tends to be easier to manifest useful and pleasant outcomes.*

Bargaining with the Divine

Do we meditate or pray or try to improve ourselves just so we can get more of what we want? Everyone cannot win the lottery. As much as I am sure God loves each and every one of us, that would entail "A Whole

Lotta Love," and at the very least there would be lawsuits over copyright infringements. We cannot avoid pain, suffering, or death by attempting to bargain with our version of God. Or can we?

Right outside Butte, Montana, there is an enormous concrete statue of the Blessed Virgin. This monument is so imposing that there are flashing beacons hung on it so that low-flying planes will not crash into it. The story of its construction is interesting. The wife of a very rich man was dying of cancer. In desperation, the man prayed in proper Catholic fashion for God to save her from her fate.

Poof and presto, in his time of trouble Mother Mary appeared, whispering, "If you build it, then she will heal" (sung to the tune of "Let It Be"). The man accepted this vision and built the statue—and his wife recovered. The statue is a monument to the power of what can occur when you enter into a powerful morphic reality and play fully by the rules of the game.

21

TAKE THE GUIDED TOUR—
YOU'LL BE GLAD YOU DID

At each level of the "reality game," we draw to ourselves the individuals and experiences we need. At each level of mastery we are currently working on, we get assigned helper energies. Those individuals or benign patterns of energy have graduated from our particular level and are available to help us with the "cheat code for our virtual reality game." These individual spirit entities or *personalized holographic helper projections* could be, for us, what we personally experience as Angels or "spirit guides." And they might look like Archangel Gabriel, your grandmother, Louis Pasteur, or perhaps even an Ewok!

I am not suggesting that whatever you believe about religious or spiritual subjects is untrue or somehow incomplete. This is just another way to look at these concepts. It is my hope that some of what I say may be helpful to many of you reading these words. When you can grasp the basic elements of these ideas, *you can begin to cocreate new and magical worlds of possibility*. Whatever you have believed up to this point, if it serves you, I would hold fast to it.

However, if you have been drifting away from traditional religious and scientific interpretations, then perhaps some of these ideas may be of service to you. I believe that Angels and Masters, saints, and other entities exist apart from me and without any input from my energy. And yet I also have come to suspect that how I experience these things is colored and seasoned by my cultural as well as personal concepts and biases.

If you have not had angelic intervention or a connection with spirit guides in your life up to this point, perhaps you might consider borrowing a page from the Vajrayana yogins, and *set out to create a connection*. At first you might experience this as being only in your mind; however, if you channel enough *zero-point energy* into the specialized morphic field of your intent, you can create an actual relationship with so-called non-physical beings. What you make up and hold as an image and then release to your subconscious can indeed take on a life of its own. It's a little like the Pinocchio story, where the loving ministrations of the lonely puppet master bring his creation to life.

I have great reverence and respect for my Angels and "spirit guides." When your life has been saved not once but many times by otherworldly voices, you *develop an inherent trust in such things*. If you have read my first book, *Matrix Energetics*, you are familiar with the story about how I was hit by a car at ten years old. I was suddenly knocked into the air, mostly unconscious.

As I was descending toward the hot, hard pavement, a voice sounded loud in my ear, commanding me to "slap the mat!" Responding instantly and instinctively, I performed a perfect judo move that broke my fall and saved my life. I did not as a young boy know anything about the martial arts. How did that disembodied voice know what to say that would cause me to react in a split second? And how did I know what it meant? Now do you see why I make such a big deal about trusting inner guidance?

Many years later, I was driving someone home who lived on top of a steep mountain pass. It was deep into the winter months, and my Suburban was plowing slowly uphill through a driving Montana blizzard. Suddenly the same voice that saved my life when I was ten spoke urgently to me, saying, "Get out of the car and put your ear to the ground!" Arguing with the voice, I said, "It's really cold out there! Not doing it!" Moments later the same voice repeated the previous message, but this time the urgency of the request had morphed into a command—one I dared not disobey. I got out of the car and put my ear to the frozen ground.

Immediately, I heard the sound of air hissing from a puncture in my left rear tire. That commanding presence that I could not see or identify had potentially saved my life again! How many times does something like this have to happen to you before you learn to trust it? Once was enough for me. *Expect miracles* and you will have them.

CHOOSE CAREFULLY WHEN AND HOW YOU LISTEN TO GUIDANCE

I have had really good experiences by listening to a certain inner voice I define as belonging to my guardian guide and friend. Does this mean that I think you should listen to any old voice that shows up in your head? Of course that is a silly and potentially dangerous idea. Ever stood looking out from a high perch only to hear a voice telling you to jump? I have, and I am sure many of you have as well. Is this a voice you should listen to? Most certainly not ever! So how do you tell the difference?

If you hear a voice that makes you feel loved and protected, that may be a voice to heed. Notice I did not say flatter. Flattering voices are just like the ones that tell you to jump from a high place or do other stupid or potentially hurtful things to yourself or another. If you hear voices like that, you should run, not walk, to a professional who can offer you good psychological help. Any voice that flatters you or condemns you—sometimes in the very same sentence—is not there to help or guide you. Any voice that tells you that if you don't stop doing something you will go blind or to hell or whichever comes first should be put on your list of suspect voices as well.

I suspect some members of the psychological community might read my words and politely categorize me and label me crazy; and they are, of course, welcome to think so. In fact, from the perceptual box that they have constructed, there is no other possible interpretation, and that is just perfect for them. I wouldn't want to be without my guardians and guides. I also wouldn't obey a voice that told me to cast myself down from a high place so that Angels might bear me up. I am not going to be listening to that line of illogic, and you shouldn't either!

If You Can't Trust Your Higher Guidance, "Hire" Another One

Dr. Mark Dunn had guides that he mentally created and then authorized to act on his behalf. A little problem he ran into is that sometimes what his guides said was either inaccurate or just plain wrong. What do you do when an employee doesn't perform according to the specifications and requirements of his job? That's right: *you fire him.* Next, you consciously revise your needs and expectations so that the next person you hire is closer to the mark and a better match for your needs. If the voice of your "higher self" is consistently wrong or just plain ornery, fire it and hire another self.

If you asked Mark or other students of mine, they might tell you they have fired numerous guides that were not measuring up. *Fire their sorry ass and start over!* Keep revising your needs and expectations. You might want to make a list of what qualities you need and expect from a "spirit helper." If you are just making all this up, then you can do anything you want. And if spirit guides are real, then it makes even more sense to hold auditions for just the right one.

Building Rapport with Inner Guidance Exercise

One really easy way to do this is to practice asking simple questions to which the answers can readily be ascertained. I am sure most of us have had the experience of getting stuck in rush-hour traffic on the way home from work. I know that you've sometimes scolded yourself for not listening to the prompting of your inner voice, the one that told you to take an alternate route home that particular day. Why not begin prompting that guide to be a more active part of your life on a regular basis? When you get in the car to go somewhere, ask your guide, "Which way should I go home?" Present clear choices, one at a time, and then pause and listen or feel for an affirmative response. If you're not sure, ask again and be patient.

Always act on the advice that is given, even if it occasionally turns out to be incorrect. You are building the strands of a strong intuitive bond, and like the Wicked Witch says in the *Wizard of Oz*, *"These things must be done delicately!"* If it helps to go out and buy some ruby slippers, or perhaps some green tennis shoes like the ones George Harrison wore on the rooftop scenes in the movie *Let It Be*, then by all means do that! Just remember one thing: this is not so much about props as about being in the *proper state* to receive the desired information.

You already know what to do if the information turns out to be consistently wrong, don't you? That's right: "You are fired!" Bring on the next applicant for the position of guide. If you fire enough of the voices in your head, the cream will eventually rise to the top. Of course, you can always have a grievance meeting with your guide if you are not quite ready to fire it outright. Try writing down all the things you want and expect from a guide so that when you have a meeting of the minds, you are clear and on purpose. One other rule of thumb I suggest: if the information you receive is cryptic or couched in symbols, be more inclined to trust it. The guides often communicate through the symbolic language of the right brain.

HEARING THE VOICE OF OPPORTUNITY

There was a young man in my seminar in Vancouver, B.C., last year whom I pulled out of the audience at random and brought onstage. When I got him up there, he told the audience and me the most amazing thing. He said that he works as a computer programmer in London, and one day he was sitting at his station when he heard an unfamiliar and powerful voice tell him he must learn Matrix Energetics.

Intrigued, he performed a search on the Internet and quickly came to the Matrix Energetics website. He stayed on the site for several hours, looking at the video clips and reading everything he could. Yet it made no sense to him. He was a software expert, and all this stuff seemed to be about healing and transformation. "Yeah, right," he thought, "whatever *that* means." Still, something there *spoke to him*—perhaps literally.

Unable to get that voice out of his head and on sheer faith, he booked a flight to Vancouver. At this point in his story, I asked him if he had ever made one of those telephones out of two cans and a string when he was little. "Sure," he replied. This was all the opening I required. Quickly, I had him hold an imaginary tin can to his ear while I threaded the string to the other can and walked out deep into the audience.

When I got far enough out, I spoke to him in a deep, powerful voice from my end of our makeshift communication device, saying, "This is the voice of God speaking." He fell off his chair onto the stage floor in a very deep altered state and remained there for the better part of an hour. When he came back to an ordinary, consensus reality, he simply was not the same man. He loved the entire seminar so much that it was almost as though he were a fish that had been forced to breathe surface air all his life and was suddenly immersed into water by some beneficent force. It was like coming home for him.

CRANK UP THE VOLUME

If you already hear the voice of your intuition speaking—and most of us do—how do you maximize that ability? It seems obvious to me that if you hear what the Bible has called *the still, small voice speaking within,* the easiest thing to do is turn the voice up! Have you ever been out cruising in your car with the radio turned down low just to keep you company? As you are driving along, the familiar faint strains of a tune break in on your meditative reverie. With a rush of happy recognition, you reach down and turn the volume knob up—way up.

DEVELOPING YOUR INTUITION STEREO SYSTEM EXERCISE

I wonder if any of you have had a moment when you start thinking about some great old song you haven't heard in years. Acting on a whim, you reach down and turn the radio on, only to discover that you have joined the exact song you were hearing in your head, in progress. This is how you can boost your clairaudience. For some of you, you might need to *look down and to the left or the right* as you find the "On" button for your intuition stereo, flip it on, and then turn it up. I'll wait right now while you do this for yourself. Do it now!

Some of you reading this may have gotten so busy in life that you have temporarily forgotten how to listen. If you are one of those people, this might take a few extra minutes of work, but it will be well worth it. I want you to read this paragraph and then put the book down, close your eyes, and slip down into a really relaxed state. When you are there, I want you to *look carefully at your intuition stereo system*. Do the simplest things first: look down and make sure it is plugged in!

Now find the "On" switch and turn it on. If it is working, *find the guide or the Angel station*. If you don't know what channel the Angels are on, refer to the program guide in your subconscious. It will help you find the easy-listening station. Make sure that when you locate your preferred channel, you mark it as a favorite so it is very easy to tune in to. How do you know if it is the right station? *You'll know*. Not because you will always hear only what you want, but because even the unfamiliar music will make you *feel so good* when you listen to it that you won't want to change the channel.

TROUBLESHOOTING TIPS

1. If you've done everything I just suggested and you are still not getting a clear signal, go back inside your mind and look at your speakers for your intuition stereo system. Are the speaker wires connected? Check carefully to make sure there are no breaks or frays in the wiring and that the other end of the wire is properly connected to the speaker terminals on the stereo.
2. If all the connections look good and you are plugged in, then perhaps you have a problem with reception in your area of expertise. For some of us, we take our jobs so seriously that we have forgotten how to play. If you think that you may be one of those people who has a reception problem, or a problem being receptive, follow steps 3 and 4 closely.
3. If reception is a problem for you sometimes, it helps to *lie down* and temporarily *forget about your troubles* for a short time. Perhaps the work channel competes with the signal from your intuition

station. Don't work at this. Relax *further down*, letting yourself sink deep inside the last time you really had fun.

4. If you can't remember the last time you had any fun, perhaps your basic life package needs upgrading. Go see a silly movie. Read a fun book or even start up a garage band!

5. If you're not the type who remembers how to have fun, put this book down and run away from it. Do it now. I'll wait. Are you gone yet? No? Well, I guess you have decided to have fun in spite of what those imperious adult-sounding voices in your head keep telling you!

6. I want you to find that voice in your head that is always telling you what to do—you know, the one that says things like "Eat your vegetables," "Comb your hair," and "You're not going out of the house dressed like that, young lady!" Or my personal favorite, "Because I said so!"

Move those voices—wherever they are, whatever they sound like—off to the side or, better yet, behind you, way behind you, now shrinking into a tiny dot in the distance.

There. That's better, isn't it?

A MATRIX ENERGETICS PRACTITIONER'S STORY ABOUT ACCESSING INNER GUIDANCE

One of many amazing experiences/shifts in my life that have transpired since encountering Matrix Energetics: I recently awoke in the middle of the night in a full panic that, at the time, felt crippling. I began to worry about everything possible: fears about money, life, school, and health. I was panicking about any and every thing that could be stressed about. I had no idea why I was panicking. I knew better than this, and yet the more I fought it, the more my body tightened and the more stressed I got.

Then around 5:30 AM I heard a voice say, "Go to the gym." Yeah, right. It was 5:30 in the morning! There was no way I was going to the gym. And I knew how to split myself into a parallel so that "other" part of me could go work out. I needed to stay put and figure this "stress" thing out. I heard the voice again firmly say, "GO TO THE GYM!" I replied angrily, "Fine. You win!" I got up and went to the gym.

When I got there, I saw some of my friends who were early-morning trainers. I smiled, said hello, and waved. They all acted like they didn't see me, like I was invisible. I shrugged it off and looked for a machine to run on. I headed for the usual machine I would run on but my body kept on walking (like I had no control) until I got to a new piece of cardio equipment. I set my heart rate monitor watch (which showed heart rate and calories burned) and started running.

Within five minutes of working out I was drenched. I'm in shape and I never sweat that much so quickly. The next moment a voice inside my head said, "Close your eyes." Intuitively, I knew I was about to learn and experience something new. I did what I was told.

I closed my eyes and heard, "Draw a hologram of yourself." Before the voice was even done saying those words, a hologram of me was right in front of me, in the center of what seemed like a black sky. Walking in front of my vision and close enough to touch it, I saw white light, the dust of glitter, and something swaying. It was another me—my higher self, my heart self (whatever you want to call it). My higher self was working on the hologram of me!

The next thing I knew it was reaching out to the hologram, pulling cords off, clearing it, and noticing what it noticed. Then from behind the hologram, from the depths of the dark sky, came numbers (frequencies), and they were flying toward the hologram. Parts of the hologram would light up, and that's where the frequencies would go and do what they needed to do.

Then out of nowhere some of my guides (Grandma, Einstein, my dog, Jake, Jesus, Buddha, and Angels) appeared, and I heard them say, "Let's play." It felt like we all sat around the round table. In the center was the hologram of me. Around the whole table was everyone else, including my higher self, and we were all there to play. In the hologram I was shown that there was no difference between them and me; we were all equal because we all came from the source of love.

Then I heard the voice again, saying, "Find peace." That's when I felt that my legs were still running, and my physical arm moved quickly behind me. I could feel my arm reaching much farther than physically possible and I grabbed an "object" that felt like peace. I brought it in and dropped it directly over the hologram. The moment I released it, the hologram fell from its stationary form in the dark sky. It disappeared, and everything and everyone was gone.

For the first time since I had closed them, my eyes opened, and as they did so, it was like I took the first breath of my life. I felt such an ease of peace and comfort. I was not sure what had just happened. I looked at the machine and saw it read forty minutes. I couldn't believe it! Then I looked at my watch (heart rate monitor) for further proof and it told me I had burned eight hundred calories in forty minutes! That is not possible, so I reviewed my heart rate monitor record captured through the workout. It was all over the place: 105 to 190 to 210 . . . on and on it went.

I got off the machine looking for a friend to whom I could explain what had just happened, but again my friends acted like they could not see me. I called people on my phone, and they had to stop me from talking and apologized because all they could hear was "blah, blah, blah, blah . . ."

I am truly honored to have experienced such an incredible awakening! Someone once asked me if I could go back and feel it again, and I said, "Yup, it is right here." And without trying, my hand moved toward a shape off to the left of my body, grabbed on, and brought me full peace and openness.

—TS, *Matrix Energetics Certified Master Practitioner*

22

THE LITTLE BOOK OF MARK'S BIG ADVENTURES

By now you are familiar with Dr. Mark Dunn, my best friend and former teaching partner for the Matrix Energetics seminars; you also know about his propensity for misadventures. He had a less-than-auspicious beginning in his career as a professional clairvoyant. Mark sees things that the average individual normally cannot and is often amazingly accurate. But it was not always the case.

Dr. Dunn's Journey into Mastery: What a Long Strange Trip It's Been

When Mark began studying with me more than ten years ago, he, of his own admission, could not see, hear, or feel anything beyond the normal purview of the territory of the five senses. He used to get exasperated with me when I would come up with some piece of clinical information that was not obtained through the normal avenues of medical information gathering. He would jump up and exclaim (the conscious mind's objection to information that is obtained through unconventional channels or methods), "*How do you know that? Don't try to tell me you 'heard it' because that just makes me crazy!*"

In order to help him have all the advantages I could provide for him, at one point I actually put a boom box under the chair where he sat in the

room with me every day. I played a cassette tape of subliminal suggestions buried deep within a sonic landscape of surf sounds and pink noise.
I don't know if that actually helped, but if you like the idea of suggestion
for accessing altered states, *you can pretend you are listening to this same
cassette whenever you read this book.*

At one turning point in his evolving awareness, Mark and I were attending a basic course on the beginning process of Shamanic Journeying, or
inner travel. Mark had amazing mental encounters with Spike, the vampire
from the *Buffy the Vampire Slayer* series, Merlin the Magician, and Jabba the
Hutt in a jeans store! Either something wasn't quite right with that boy's
brain or he was beginning to show tantalizing initial glimpses into the complex and amazing abilities that he demonstrates so easily today on a daily
basis (such as accessing archetypal and dream imagery). Based on what I
know about this incredible individual, I'll go with the latter interpretation!

Interestingly, the person who taught this basic Journeying course
admitted to the class that at first she couldn't Journey at all! When she
was first exposed to shamanism, she struggled daily for two years before
she had a single successful Journey. She just refused to give up. *There is
no such thing as failure in the process of developing your intuitive capabilities. Everything is just results and information.* Persist in a relaxed
manner and you will develop your own extrasensory abilities. Key points
to developing any new ability: Relax and let go. Do not judge. Mentally
create a safe environment where you can explore altered states and paranormal abilities.

MARK'S FIRST INNER GUIDE

Shortly after attending that first Shamanic course, Mark received his first
"inner guide": a cartoon moose—and no, it wasn't the famous one! This
guide told Mark some apparently useful things from time to time. There
was only one minor problem. Most of the stuff Mark heard with his
"inner ear" was initially wrong, according to his clients' feedback.

I repeat: There is no such thing as failure. There are only results and
information. The antics of Mark's guidance were in keeping with a song

my professor at Bastyr once wrote called "My Guide Clyde." In the song, Clyde kept telling the person all these things that in retrospect turned out to be completely wrong. In the final verse the protagonist finally gets around to *asking the key question*: How come many of the things you've told me turned out to be false? Well, Clyde the Guide was an alcoholic and was just making stuff up. It is good to test the spirits to make sure that they are not habitually imbibing them! Developing the spirit of discrimination is one of the first tests you must pass on the journey to spiritual enlightenment. This teaching in some form is emphasized in all the Mystery School traditions I am familiar with.

Always question the information you receive through any and all means. Never become complacent in this process and it will serve you well, adding new dimensions into your experiences in this life. With that being said, let me caution you not to ever become discouraged or to give up. And never assume that you know anything or that what you view as reality represents an absolute truth. As you grow in your ability to perceive in a new or fresh way, your perceptions of what is true for you will frequently morph and change.

About the same time the cartoon moose showed up as Mark's guide, another phenomenon also began occurring in his world. Sometimes when Dr. Dunn was talking to a client, he would see intermittent blue bubbles or spheres interacting with some part of the client's anatomy. Wisely not trusting his initial psychic impressions, if he saw a blue bubble sitting on a client's shoulder, he would ask the client if he or she had a problem with that particular limb. More often than not, the client would often exclaim, "*No!*" It got better over time—much better! Notice what you notice, but never assume that the information is real or ultimately important. In other words, judge ye righteous judgments. What that means is follow the energy in the moment.

After putting up with the "blue dot phenomenon" for a time, Mark decided he would try priming the pump to increase his accuracy. One day I walked into his office area and discovered him huddled over an anatomy chart with a pendulum dangling loosely between his index and middle finger. He was patiently posing yes and no questions about his next client before he physically arrived in the office for his appointment. *Find or seek out your own ways of obtaining personalized intuitive information.* Some people use muscle testing, pendulums, dream imagery, lucid dreaming, shamanic journeying, meditation, and other methods. Explore and find out what works for you. This proved to be a process involving a lot of trial and error, but gradually over a period of a year, his accuracy began to steadily improve. Since Mark always verified whatever information he got intuitively, there was no danger or shame in trying to predict things about his clients. This provided a conceptual safe haven in which his talents steadily grew and blossomed. Trust but verify. Never assume psychic information is accurate or correct. Constantly seek verification through some means, and strive to continually improve your accuracy in simple and useful ways.

ADVANCED LESSONS IN DUALITY

Maybe you don't quite believe in Angels. My dear friend Mark wasn't sure about them either. After all, he had never seen one and certainly could not openly claim that they had ever saved his life. Mark had heard all the stories but had never consciously experienced an Angel. He prayed and meditated and cajoled, but no Angels appeared. So one day he decided to make his own. Having chosen a course of action, he researched everything he could find about Angels on the internet. In his research, he found the most information about Archangel Gabriel, so he decided he would model his own personal Angel on him.

By engaging in a deep meditative process, Mark interacted with the morphic field of the universe. Focusing intently, he attempted to "channel" angelic patterns of energy into a mentally created template. When this virtual reality construct was completed, it gave all the appearances of being alive. You heard right! Mark could walk and talk with his version of Archangel Gabriel.

I didn't know it at the time, but there actually exists a well-established precedent among a particular spiritual tradition. There is a Hindu sect of Vajrayana yogins who have a practice similar to what Mark intuitively did on his own. They memorize in exacting detail a chosen god form. Over time, the practitioner, through deep meditation and concentration on the chosen form, develops the ability to interact with it in a manner similar to a virtual reality simulation. Eventually, the aspirant of this amazing discipline will be able to see the form as clearly as a table or chair in a room and to interact with it as well. Michael Talbot talks about this process in his book *Mysticism and the New Physics*.

There Must Have Been Some Magic in Those Old Wings He Found

(Sung to the tune of "Frosty the Snowman")

Mark intuitively hit upon the duality approach, it seems. One day he was working in the office and a client came in having an acute gallbladder attack. The patient was doubled over from the severe pain. This condition can constitute a true medical emergency. Consistent with his duty as a primary portal of entry medical provider, Mark informed her that she should go to the emergency room. She, however, asked if he could try to do something before she exercised that option, so he agreed. Twenty minutes later the woman was still in a lot of discomfort and Mark had exhausted every possibility in his medical bag of tricks and techniques— or so he thought.

Gabriel Gets Real

Looking around the room for anything that might help, his gaze landed on his version of Archangel Gabriel. His virtual Angel was at that moment, of all things, appearing to be eating a peanut butter and banana sandwich! In exasperation Mark addressed the Angel, saying, "Can't you do anything about this?" Slowly getting up from a seated position, still chewing, the Angel walked over to the client and calmly thrust his noncorporeal hand into her abdomen. Instantly the woman's pain completely disappeared. "What did you do?" she asked, shocked by the sudden reversal of her symptoms. Mark, feeling a bit ill, told her, "Believe me, ma'am, you really don't want to know!" With that, Mark impressed upon her the need to still go to the emergency room to be safe. She complied with his request, but for all intents and purposes the crisis was resolved.

I don't have to tell you that for a time after that everything was all about Angels for Mark. In fact, the day after that experience with the woman with the "hot" gallbladder, I went into Mark's room to find him in a meditative position head down on the floor in a posture of reverent awe. I could vaguely see the outline of two brilliant forms standing in front of him. "Sshhh, I am communing with my Angels," he informed me. I nodded in silent understanding and gently closed the door behind me on my way out.

IN LIGHT OF ALL THIS, NOW WE VISIT THE DARK SIDE

This was Mark's beginning and intense crash course on lessons in duality. Surely it is not hard for you to puzzle out. If you set up rules in your unconscious so that you polarize or divide everything into opposites, what do you get on the other side of Angels—the wrong side of the light tracks, shall we say? That's right. You win a prize. You can just see Bob Barker emceeing some demented otherworldly game of *Truth or Consequences*. If there are Angels in a relative sense, in the expression of that duality there must be Demons. Ta da, tat ta da! (Cue fanfare!) Lions and Tigers and Demons, oh my!

Mark started seeing possessing spirits in every client he worked with. For a time his medical practice looked a lot more like an exorcism than anything taught in medical school. He took this phase in his learning to such a degree that at one point he was asked by another self-styled spiritual warrior to help him exorcise a church of possessing spirits. Who needs to go to the movies when your life starts to look like this?

For a number of months, Mark engaged in the clinical equivalence of spiritual warfare. In primitive cultures and shamanic thought, possessing spirits can be seen to be the cause behind the manifestation of any disease pattern or condition. The New Testament displays the prevalence of this idea in many instances where Jesus performed a healing by casting out unclean spirits. Mark for a time adopted a similar or shamanic model of reality. Everywhere he looked there were possessing spirits!

Eventually, lugging around an imaginary sword with which to do battle with invisible forces loses the force of its charm. And, oh yeah, you can start to go a little nuts! Remember what I have said in other parts of this book: the concept of superposition allows for the state of all possibilities to exist. As soon as you choose a particular perceptual model for reality, you will look at life in all its multivaried aspects through the lens of beliefs that you have chosen to adopt. You will see what you expect to see within the constraints of your individual or cultural model.

If you don't like what you are seeing or experiencing in your life, changing what you look for and expect is an excellent way to begin to mold your life into a more desirable set of experiences. Mark finally realized that his perceptions could be viewed like a kaleidoscope. If he did not like what he was seeing, or if seeing things in that particular manner was not inherently helpful, he could change the scope of his perception. As with a kaleidoscope, when you twist the toy's mechanism, different patterns appear before your very eyes.

When Mark stopped expecting a particular outcome and relaxed, he began to expect that whatever he would see or encounter was uniquely useful in the moment. If he found himself in an uncomfortable or less-than-useful reality, he adopted the habit of asking the question, "If I were to see something that could really help this individual the most in this moment, what would that be?" Then he would center his awareness gently in his heart and wait for what would present itself. By doing this, he let go of the need to judge or analyze the information, which changed the nature and character of the entire encounter.

This was probably one of the most powerful lessons I learned from Mark. If you don't like what your experience in life is showing you, drop down into the sacred space of your heart, and from this vantage point ask, "If I were to begin to have a new and better experience in life, what would it look like or feel like?" Trust that the universe hears your inner entreaty, and see how life can change for you. Miracles are everywhere if we can just open our eyes to see them!

23

CREATING ADVANCED ARTIFICIAL TEMPLATES

My friend and mentor Dr. M. L. Rees wore a large crystal over his chest in the traditional location of what some cultures and belief systems refer to as the heart chakra.[1] He called this crystal construct a "doc harmonic." He designed this master crystal to energetically link in to a number of programmed crystalline-based devices, which he kept in his office, located in the most unlikely of places: Sedan, Kansas. The concept he created was one of programming these devices with a specific intent to heal or correct specific disease conditions or problems.

Dr. Bill Tiller, subtle energy and psychoenergetics expert, talks about imprinting very simple electrical devices with a specifically programmed mental intent to perform very specific tasks—such as adapting the pH in a room or changing other physical parameters that are objectively verifiable and can be measured through scientific instrumentation. He wrote about his experiments in a wonderful paradigm-shifting book titled *Some Experiments in Science with Real Magic.*

Tiller looks a lot like some people's archetype for a wise wizard. Do you suppose it could be because he embarked on a course in personal self-mastery more than thirty years ago, which included meditation, concentration, and yoga? Tiller is one of the physicists who assessed the claims of Russian scientists concerning psychic abilities during the years of the Cold War.

Many of the scientists who watched Uri Geller and Russian psychics perform feats involving apparent psychic prowess proclaimed that what was being observed was the result of sleight of hand and sideshow trickery. Tiller was one of the few dissenting voices that spoke up and proclaimed that he thought there was something to what he was observing.

Tiller reasoned that perhaps we had not developed measuring equipment sensitive enough to be able to assess what was "really going on." He further reasoned that it must be possible to develop these attributes to some extent, so he resolved to fine-tune his sensory mechanisms by embarking on a journey of self-discovery and spiritual discipline.

Over the course of years, Tiller formulated an evolving set of hypotheses that strived to account for the phenomena he witnessed so many years ago in that government lab. His journey of personal discovery and self-mastery continues to this day.

Rees was very scientifically minded; he was even part of the engineering team that developed the little black boxes that were fitted into Allied planes in World War II. These devices would emit a signal to our forces on the ground that the plane was identifiable as friend rather than foe. Rees then studied osteopathy and chiropractic, earning degrees in both disciplines. But he did not stop there. Over a period of forty years, he developed many creative innovations and techniques that other doctors learned in order to better serve suffering humanity. Some of the things Rees discovered and developed bordered on and even crossed over into what would appear to be Magical Realms.

In certain ways Tiller and Rees are very much alike: both are pioneers, always seeking answers that reach beyond the thinking of our current limiting mind-set. Just like Tiller, Rees knew that there were many things he was experiencing that could not be explained by his engineering or medical backgrounds.

Resolved to make a difference, he set upon his own course of self-discovery. This led him to the secrets and traditions of many cultures concerning the hidden side of life. He confided to me that after years of frustration and much personal effort, he finally had a personal breakthrough where he could actually see the energies and forces he was working with. He used his abilities garnered over many years to heal and bring comfort to all who came to him for help.

Rees personally trained me in the use of his crystal-based technology called "Harmonics." In addition, he personally inducted me into the order of healers that he took under his wing for training. This organization was called ISHO, which stood for the International Systemic Health Organization. It was dedicated to the healing of the ills and diseases of mankind. Rees once told me that he could "go into any hospital, and if given ten minutes with each patient, [he] could clear out the facility." I cannot personally attest to the validity of that particular claim, but during the time I was privileged to study with him, I witnessed and accomplished some amazing things.

For example, my first son, Nate, was born with bronchitis, asthma, and allergies. He was prone to frequent bouts of pneumonia, which took hold of his frail little body with regularity every couple of months (as detailed in my first book, *Matrix Energetics*). Neither conventional medicine nor alternative approaches addressed his particular needs, so I tried to learn everything I could that might provide a solution, or at least help. I read every weird book and studied every strange technique I could get my hands on. Thankfully, my chiropractic school's library was a compendium of unusual and downright arcane knowledge. It was there that I first encountered Rees's book *The Art and Practice of Chiropractic*. Unable to decipher much of anything the book was referring to, I called Dr. Rees's office and, to my surprise, he personally answered the phone. I inquired if I could ask him some questions. He replied, "It's your dime, young man." I then said, "Dr. Rees, I did not expect to get you on the phone. I am reading your book, and it seems like you are from outer space." He laughed softly, "Well, that could be, young man."

I spoke to him for a while, and then he made me an offer. "I tell you what. If you can get a few students together, I will come out and teach you guys." This conversation led me to invite him to come to Dallas to give a seminar to a select few students who had interests similar to my own. The chiropractic college was progressive, but not that progressive, so we assembled at the annex of the school, aka the nearby Holiday Inn.

Rees created the system he called Harmonics. He looked all over the world to find precious gemstones and various other things that could be used as a repository for healing energies. These harmonic constructs could be shaped and molded into powerful artificial templates for healing. Although Rees did not call it such, I believe this technology was based upon principles of scalar physics, which I've referred to in other parts of this book. These things looked like little plastic disks, and he would put various crystals within them at certain key geometric positions.

He would then solder them together so that they formed a kind of circuit. Then he put metal flakes over clear plastic resin to act as an antenna to pull in information from the aether. He would then smooth over the top of the resin, allowing it to harden. To complete the device, he would inscribe informational symbols on the face of the implement. These engineered templates could then be activated by the doctor's focused will.

Then he tied certain spiritual forces to it and would metaphorically insert and activate it into the person it was created for. Obviously, trust is the foundational principle here. He actually made hundreds of these disks. At one time I owned a modest set of them. He had one for minerals, one for viruses, and so on. Once he had hundreds of these, he programmed them into a matrix in his office. Then he passed out "Doc Harmonic" master crystals, and the doctors open to this system would just call up the energy they needed into their cupped right hand and people's diseases would go away.

If you create something like that based on the engineering principles of a paleo-ancient science and spiritual technology, it can be incredibly powerful!

Because I was seeking the Holy Grail of healing techniques, my brown leather medical bag did not hold the traditional medical implements. Instead, I had a "double-headed" adjusting instrument, a plastic case full of strangely labeled glass vials, and a few of Rees's Harmonic crystal technology "thingies." These objects became the symbols for my awakening into new dimensions of healing technology and knowledge.

In one of my student clinic shifts at the college, a young woman came in for treatment and for some reason chose me as her intern. When I took her medical history, she confided to me that she had not had a period in eight months but wasn't pregnant. Since she was only twenty-five years old, this was possibly medically significant.

After I performed an overly thorough physical, with the exception of a gynecological exam, my thoughts turned to how I might be able to help her. I had her lie faceup on the adjusting table in the exam room and rooted around in my medical bag for clinical inspiration. My hand was drawn to one of Doc Rees's harmonic contraptions. This particular device had a bright turquoise plastic base in which was set a large quartz crystal shaped like a pyramid. Not knowing what else to do, and having limited clinical options, I held the base of the crystal device and pointed the apex of the crystal in the direction of the woman's lower abdomen.

Developing a vague glimmer of a plan of action, I moved the device in smooth, slow circles above the patient. This activity went on for a couple of minutes, and then the lady began to writhe and move on the table. I wasn't sure what she was experiencing, but if I had to guess from her facial expression and body language, it wasn't entirely unpleasant. As this development came to its peak activity, she suddenly informed me that her period had just started right there on the table. I had no idea what had just happened, but I was rapidly developing a deepening respect for weird medical approaches.

I did not hear from that particular patient again. Six months later, however, I got a call in my new office in Montana, where I had moved after graduating from chiropractic school. The caller on the other end of the line informed me that he was a member of the Texas Chiropractic Board and that I was in trouble. It turned out that the doctor I was speaking with was the boyfriend of my patient who had the miraculous menstrual cure—and he was not amused with me.

He could not even begin to deal with the weirder aspects of his girl-friend's experience, so he focused on something he could understand. He accused me of practicing medicine without a license, intimating that he was thinking about turning me over to the Texas Attorney General's office for criminal prosecution. Perhaps with an echo of the personality I would later take full possession of, I asked him what grounds he had for the accusation, knowing full well that he wasn't about to discuss crystals and weird implements.

I was right. He focused like a pit bull on the fact that I had told his girlfriend that I could adjust her lower back vertebrae to free the nerves of her pelvis from impingement. The L5 (lumbar five) nerve supplies the uterus, via the pudendal plexus and other tributaries. His contention was that by adjusting L5 to help uterine function, I was practicing medi-cine without a license. "What an 'ass-set' this guy is to his profession," I thought.

I had had enough of his passive-aggressive posturing. I told him that in a book called *The Science, Art and Philosophy of Chiropractic* by D. D. Palmer, it was specifically stated that too much or too little nerve force to an area could cause disease. Further, through adjustments to the verte-brae in order to release the neural impingements, the life force and flow of information to the area supplied by the affected nerves could be released, thus restoring the target organs and tissues to proper function-ing health.

"Have you ever heard of D. D. Palmer?" (the founder of Chiroprac-tic) I roared. I then slammed down the phone, never to hear from him again. I guess he went back to his college textbooks and looked it up. Maybe his girlfriend's new reference for effective chiropractic care had become weird crystal devices and he couldn't deal with that.

If you think that this was an isolated event with this technology, I will share a few more stories that may convince you otherwise. The first one that comes to mind is about the man who came to me with such an acutely painful lower back condition that there was nothing I could do within the narrow parameters of "normal" chiropractic practice. Resolved to wing it, I reached into Dr. Rees's bag of tricks and settled upon a brightly colored plastic doodad. This one was almost a fluorescent blue color and had four small quartz pyramids embedded in the plastic base. "This has gotta do something," I thought.

I led the suffering patient into a quiet, dark room in the back of the office and had him lie facedown on one of my adjusting tables. I put the device flat on top of the man's lower back, turned off the lights, and left the room. Forty-five minutes later I checked in on him and was shocked when he announced that all the pain was completely gone! As proof of this startling statement, he stood up, no longer bent over double in agony. A huge smile was painted across his face.

Reaching down, he grasped my hand gratefully and said, "Thanks, doc. You are a miracle worker!" I was quick to assure him that I did nothing and had no idea why or how he was so much better. Later, when I told Rees this story, he laughed heartily in amazement. The device that I had used was designed to work like the trap used by the Ghostbusters. Talk about feeling like *The Man Who Knew Too Little*. I had no clue, but it had worked anyway. Rees said, "He must have had a 'hitchhiker' causing the trouble."

One last weird fact in this case: when the patient stood up after being treated, his breath had a fruity alcoholic odor to it, which he did not have when he came in. That smell is one I associate with diabetics whose bodies are producing too many ketones. This suggested that a biochemical change had also occurred in the process.

One other vignette about Rees technology concerns a man who lived with chronic, intractable pain in his heels and the soles of his feet. Emboldened by my previous successes, I put another one of the strange little devices on the client's lower back and had him lie down to relax. Sure enough, after about twenty minutes he announced to me that he was feeling quite strange, as if he were floating above his body. I had him stand up and walk across the room. Both the client and I were overjoyed to discover that the previous chronic pain in his feet had melted away into nothing.

Rees would assign "spiritual beings" to the software after he decided what each one was to accomplish. He would then assign the energy of the beings to the little points anchored by the gems. This then becomes an engineered interdimensional matrix or manifold that accesses hyperspace dimensions.

A problem arose however in that Rees's work started to feel alien and somehow disturbing. I became a bit distrustful and felt like I was not supposed to use that technology anymore, so I stopped. Then one day a strange thing happened.

I was traveling with a very psychic friend of mine in Livingston, Montana, and it was a cloudy day. There was a big circular cloud that seemed to be following our car. My psychic friend concluded that it was an alien grey ship. He said that I had something they were interested in. I had a device from Rees with me that was programmed with spiritual and energetic technologies. So I handed it to him. He looked at me and said, "Very cool, Richard. I am talking to the captain of the alien grey ship above us, and he says he likes your technology." I responded, "You keep it."

Many years later the Angelic intelligences that sometimes advise me indicated that it was time to re-create a new version of Rees's technology. Accordingly, I was given new detailed instructions concerning how to create a kind of energetic software that some might call imaginary. This Angel (or whatever it was) said, "We want you to call them Modules, and we will guide you step by step in their construction." Okay. There is that trust thing again: either you do trust or you do not. Mark and I decided that trust was the better approach, and we spent some serious hours creating what the Angels had beckoned.

What Mark and I actually wound up doing was assigning sacred geometry patterns merged with Kabalistic knowledge and quantum physics patterns, arranged in grids. We took that as our starting template. We had books on minerals and gems with nice color pictures and descriptions of everything, and we went through the books clairvoyantly until we were told to use a particular one.

When we had enough components for the construction of a particular Module, we would then ask where they needed to be placed in the template we were building. Working with the virtual forms of Angels and other virtual beings, we would ask questions such as "Who wants to play on this point on the diamond?" These forces were engineered from the energy of the vacuum and placed strategically within a desired form in order to hold the software instructions for our design.

In essence, we developed an interdimensional technology that holds a space from the zero-point field and creates a conceptual bridge point from the imaginary to the real. This describes, by the way, a complex conjugate number in quantum physics where minus one represents the imaginary quality, and then there is the real number, or quality, as well. When you multiply the two together, the imaginary component cancels out, and you end up with a real vector or coordinate in space-time. This is exactly how this energetic technology works. It is a spiritual technology that is engineerable and reproducible.

Here is how Bob, one of my students, creates a Module. Bob is an RN in a hospital and freely plays with Matrix Energetics and Modules, such that he is now often referred to as the Matrix Magician. He reaches into a dimensional window and pulls out what he feels, trusting that the exact thing that he needs will be there waiting. Speaking from the true authority of his magnanimous heart center, Bob simply says, "Create a Module for this condition, thing, or circumstance." Grasping the newly created virtual technology in his right hand, he activates it and then installs it wherever his perceptual awareness guides him. Nothing could be simpler. What is his first technique? TRUST.

CREATE A MODULE EXERCISE

So this is Bob's process: *Create*—Reach into your mind's creative window. *Wait and trust*—You will actually feel something building in your hand. If you do not, then trust that you did. *Install*—When you say install, you will see, or know, where it needs to go. *Activate*—Now it has that intelligence in it, and it is ready to do things. It is that easy.

Bob's process rocks! It is so simple and yet so powerful. You might as well use stuff that is working really well. You do not want to use stuff that does not work so well.

My guides told me that there are quantum Angels from the seventeenth dimension that make this stuff. Whether you believe that or not is immaterial to me. After teaching so many seminar participants about the Modules, the morphic field of their reality has grown large and powerful indeed!

You do not have to have a medical mind-set, and if you do not have any reason to do this, then do not. But it is kind of fun. The real point of doing this is to play and trust. If you can learn to do this, then if your life is at stake, you will have already established momentum. In this way, you can build the muscles of faith in things hoped for but not yet seen.

You may choose to construct Modules the way I do—the mental way—or you can just trust. The latter may be a better way. To summarize again, here is what you do: *Create* the module, *Activate it*, and *Install it*. That is all there is to it. Trust that there is information between the spaces. Really, you do not need to know anything. You can simply trust that it is already done, and you will receive the best outcome. Trust is going to be your most useful tool. So if you have issues with trust, you might want to create a module called **trust** to support you. It is very useful.

How you build your module is based upon your own individual knowledge and intuition. Some of you may say, "I cannot do this" and then you will realize you can. Others will say, "I can do this," and then realize maybe not. Still others may say, "I can, but maybe with a little work." You could also say to yourself, "I want a module for understanding what Dr. Bartlett was talking about."

Here is a recent story from a student in Matrix Energetics:

Yesterday my son-in-law's father was taken to the hospital. He was in excruciating pain with swelling in his back where he'd had surgery two weeks ago. When they arrived at the hospital, the family was told that it was most likely a blood clot, especially since it was on his left side. The swelling was causing the nerve pain down his legs.

When I talked with my daughter and received this news, he was on his way for an MRI. I started ME I two-pointed for what was needed. It looked to me like there was no blood clot. I imaged the swelling being some sort of fluid that would be absorbed back into his body with no ill effects.

Then there was also some sort of Module I downloaded and installed at the site of the swelling. It felt to me that it was to break up any possible congestion.

When my daughter came by later that evening, she told me that the MRI had shown only some fluid, which the doctors said would be reabsorbed by his body!

—BD

24

THE MANIFESTATION CHAPTER

$\int$ ome people are born wizards and other people have to work at it. If you have the ability to manifest things, then chances are that you will. People who have built a lot of unconscious resources can depend on the world to treat them the way they expect it to. In one unreal sense, there is no you and there is no world. Building rapport with wealth, love, or even the ability to perform healings of yourself and others requires building *coherent useful references*. If you have seen examples of good relationships, or perhaps even had a few yourself, it is much easier to know what ingredients to put into the wish's cauldron for the magic to work. *Can you say "congruence"?*

To be congruent with magical states of possibility, you must think and feel in the language of magic or little to nothing happens. I remember watching a Clint Eastwood movie called *Firefox.* Clint was a highly trained American pilot who was sent to Russia to steal its experimental thought-controlled superjet. Clint was able to steal the bird and get airborne without much difficulty. Once he was airborne, a second Russian prototype gave chase. When Clint tried to fire a missile at the pursuer, he couldn't do it until he remembered that he must think in Russian. Every form of success or endeavor has its own language and syntax.

In order to succeed you must speak and think in the language of the virtual reality game you are attempting to master. Here's a clue: the phrase "I can't" is universally *not* a cheat code in any of life's simulations. Time and again people have attended the Matrix Energetics seminars thinking that being present with me was somehow going to gift them with the key to the magical siddhis.

If being present in a seminar invokes a magical state, the trap inherent within that belief is what happens when you go home. If you just mimed the language and went through the motions of the techniques taught, that may not be enough. If you are a healer looking for another tool to add to your technique utility belt, that level of experience may work fine for you. If you wish to really master the magic, you must contact and become the inner magician.

Like Clint, if you unconsciously assimilate the language of Matrix Energetics, then you no longer have to consciously think about it. At the level I am referencing, *there is nothing to learn how to do; you have become.* The Ascended Master Saint Germain once said, "There is an alchemical formula, but by the time you discover it you have long since become it."

One of the methods of opening the way of the inner magician within you is to ask liberating questions such as "If I did fully believe in myself as the master magician, what have I become now?" There is a master within you who is always listening to what you say and how you say it. Your words, as holographic cups of light, form drop by drop the chalice of your inner mastery or misery. It is for you to choose and, having chosen, continue to choose with each breath.

THE HAVE AND HAVE NOT RECIPROCAL EQUATION

Jesus said, "For whosoever hath, to him shall be given and he shall have more abundance: but whosoever hath not, from him shall be taken away even that which he hath."[1] Why do you think he said this? This is what I think: When you are in a consciousness of "have not," you are constantly instructing your unconscious mind to notice and create more of what you are focusing on. When you habitually do this, you will find more experiences to match what you are focused on. In doing so, you prove to yourself that you are powerful in your state of seeming powerlessness. You create an attraction to what you don't want or have.

One of the easiest ways to begin manifesting a new life, with all new parameters, is begin to *drop the attractors or charge on what you don't want*. You begin to realize that there is an equation of consciousness in play here. If you have an electron, a negatively charged particle, then the balanced force is the positively charged proton. Equally, for every photon there is an anti-photon. When the charges are in balance, what you get is a back-and-forth flow, or charged polarity, which creates movement. When you are in the flow, you don't notice that there is a paired photon/anti-photon.

When you've had enough of what you *don't have*, you don't have to do things that way anymore. You can then flip into the state of *have*, which is the antithesis or antiparticle to what you don't have. The have/have not equation is just like the photon/anti-photon pair that creates the *graviton*. When you pair the have/have not state, you have a balanced charge and you don't notice the presence of duality. What you experience is the flow of the movement between the opposing charges or forces. The balanced expression of these two extremes represents the electromotive force we call Life. *Life has its ups and downs because this represents a balanced spin or state of flow.*

When you let go of the distinctions that engender a multiplicity of dualities, you have stopped resisting the flow of life. When you lower the resistance to the flow of light as information, you are invoking my version of OM's law. When you stop resisting the flow or tide of events, the current's strength and speed is magnified. When you begin manifesting a greater current, you could say your chi, or life force, becomes magnetized and magnified.

When your life force is increased, that is when you want to begin to monitor your thoughts, feelings, and unconscious momentum a little more closely. When you turn up the juice, you have more power to focus your intent into manifesting marvelous and magical changes in your life. You also can manifest what you don't want. This is a good reason why you want to maintain a state of childlike wonder and gentle neutrality.

What you manifest doesn't come from the realm of your conscious mind, in any case. If your ability to manifest what you want originated from the domain of the conscious mind, you would continually see the results of your every thought played out in your life in front of you. For most of us, we are not in an entirely positive or negative mind-set all the time. Therefore, the results would probably even out over time. If both positive and negative outcomes were evenly manifested, then the net gain could be seen to be approximately zero.

Thoughts that have no conscious direction still wield just as much force, but do not have a clearly defined destination. Since the power of our thoughts entails the release or expenditure of energy, the charge of our thoughts has to go somewhere. If energy can be neither created nor destroyed but merely changes state, where does the energy of our directionless musings and desires go? Just like the pairing of the photon/antiphoton, our thoughts are paired or grouped as conscious/unconscious.

Our unconscious thought energy therefore resides in the **morphic field**, or **quantum potential**, of the unconscious mind. Since we are dealing with the paired states of have/have not with every desire, that energy must go somewhere. If we therefore habitually focus on what we want, the energy of opposite polarity or charge, the have-not state, becomes stored potential in the unconscious field.

THE CONCEPT OF A MORPHIC FIELD DEFINED

To remind you what a morphic field is:

> Sheldrake regards the morphic fields as a universal database for both organic (living) and abstract (mental) forms, a field within and around a morphic unit, which organizes the field's characteristic structure and pattern of activity at all levels of complexity. The term morphic field *includes morphogenetic, behavioral, social, cultural, and mental fields. Morphic fields are shaped and stabilized by morphic resonance from previous similar morphic units, which were under the influence of fields of the same kind. They consequently contain a kind of cumulative memory and tend to become increasingly habitual. Swiss psychiatrist Carl Jung brought the theory of collective*

unconscious to our awareness, and with it a similar proposal that a mode of transmission of shared informational patterns and archetypes exists. According to Sheldrake, the theory of morphic fields might provide an explanation for Jung's concept.[2]

TAPPING INTO THE MORPHIC FIELD OF THE MATRIX

In the beginning of the book, I said that Matrix Energetics has a huge morphic field that allows you, with minimal effort, to step into a unified field of consciousness. This is worth repeating here since this is indeed the secret to manifestation. It is this powerful group dynamic that allows you to amplify your desires and abilities in service to the collective good of all concerned. This field already exists; it is to your benefit to simply plug in. When you tap into the morphic field of Matrix Energetics, your results can become more powerful and reliable.

CAN YOU SPOT THE FALLACIES IN YOUR THINKING?

If we invest a lot of energy in manifesting something, an equal amount of paired energy must then be stored as destructive or unmanifested potential. If we then store unconsciously that "have not" energy and attach an emotional charge of fear to those thoughts/feelings, we have, in effect, created an artificial *highly charged quantum-feeling potential*. This is like an electrical potential, which under the right conditions can leap from the ground to the clouds like a lightning bolt.

However, if you can create an artificial potential of studied neutrality with your manifestations, you begin to drain the deep swamp of unconscious fear-based patterns. When this is done over time, you free the negative charge on your human/divine potential, freeing up more energy to flow unimpeded into the manifestation of the things of your heart's desire. *Your manifestations are driven by the engine of your beliefs and your experiences.* I used to think it was really hard to change the content of my beliefs and that I had to work really hard at doing it. The idea that deep, lasting change is really hard to achieve is in itself simply another belief.

Here is how one practitioner applies Matrix Energetics to his process for manifestation:

I'll try to explain how Matrix works for me, but understand that everyone has a way of interfacing with it, and how I do Matrix may be quite different from how it works for someone else (probably because we all bring different backgrounds and perspectives to it). So you have to play with it and see how it works for you.

I rarely have a desired outcome in mind. If you only use Matrix Energetics to manifest what you specifically intend, then you limit it to being able to provide only something that you can imagine. This prevents or limits something BETTER than you can imagine from showing up. Let's take money for an example. When I two-point my finances, I place the energetic pattern of my finances in front of me. I simply intend that to happen, and I often see it as a pattern.

Usually patterns look to me like a luminescent vortex or a very convoluted Borg cube (if you're a **Star Trek** *fan). The shape itself is not important. Trusting that the pattern is in front of you, whether you can imagine it or not, is what's important. (Doing Matrix is all about trust; as Richard says, the more you trust, the more powerful you are.)*

Knowing that the pattern is in front of me, I reach out (usually physically with my hands) and find a point in the pattern that feels stuck, hard, rigid, or different. Then I search for a second point in the pattern that intensifies the feeling in the second one so that I sense that the two points are connected. Then I drop down and let the two points change into a more useful state, one that feels better than how they were at first.

This process has shown to correspond to healthier, more abundant finances. I don't know (and don't need to know) how this occurs. It may be beyond my comprehension. The more I do not know, the more powerful it seems to be. The most important thing is to develop an attitude of trust.

—EO

If you dive deep enough into your own psyche, you can attract whatever you think about—and that is no great secret. The secret is in being able to think about attractive things consistently enough that they *become magnetized* into your personal orbit of "Becoming." When you resist the flow of the *feel good and trust current*, you give life to and sustain a counterwave of energy called fear and struggle. Because when you consciously sustain a desire or want, the phase-conjugate opposite reality—exactly and precisely what you don't want—is also manifested.

This counterwave can interfere with your creative visualizations, affirmations, and wishful constructs. You are bleeding off precious spiritual and psychic energy. When you do this, the result is that you have less energy available to manifest and sustain the focus of your creative intent. *One of the best ways to manifest what you want is to stop wanting and stop manifesting.* In this way there are fewer struggles, thus freeing up far more energy to be directed unconsciously and unerringly to the target of your secret desires. Things will just show up and *people will think that you're lucky!*

THE SECRET TO MANIFESTATION LIES IN THE NOT DOING

Many of us in this culture seem to believe that in order to make something happen *we have to do more.* But the more we do, the more involved we become in the need to do, have, and be. When we need to have something in order to know that "we be alright, we be making an error." My little daughter with autism is always saying in times of emotional upheaval or crisis (yes, I do have plenty of those moments still), *"Be all right, be all right!"* This sweet voice always underscores for me the fact that the universe is all about Love. We are evolving because "evolve" is an anagram for "loving." **Love is the decisive, cohesive force in the universe.**

SUPERCHARGING YOUR MANIFESTATIONS

We all have a well-developed ability and momentum to manifest things. The question then becomes, which part of our programming runs our manifestations? Is it our conscious expectations or deeply ingrained unconscious belief patterns? Which one do you think it is? A question to ask yourself is, "Do I like what I am manifesting, and if not, how do I change direction?" If you set off into the waters of the unconscious and want to change things in your life, it helps to set the sail of your intent in the direction that the wind is blowing.

WHICH WAY IS YOUR PSYCHIC WIND BLOWING?

One time-tested way to determine the wind's direction is to wet your finger and hold it up in the air, rotating it all around you until you feel the force of the air currents on your finger. You can do this psychically as well, and you might want to stop reading, put this book down right now, and hold your index finger up in the air. Move it slowly around you in a complete circle, probing and feeling for a change in yourself or your inner state. Don't overanalyze what you feel, but look for a direction that points you toward a state of feeling really good. When you catch the drift of that feeling, psychically push off from your life's emotional sandbar and move out in the direction that the wind would blow you.

A MANIFESTATION EXERCISE

Do an experiment with me. Pick something right now, while you are reading this, that you would like to see shift in your life. Got it? Good. Mentally extend that thing out in front of you several feet or more as a single point in space. Now, without considering what the result would be, choose unconsciously a world of possibilities where that condition or thing you desire could show up differently. Now let go of the need to do anything. Feel the expansion for the state of possibility as you collapse the former reality into the new one. Now let go and accept the change that has taken place.

If you really do this and you enter into this exercise and feel it, your life will change. I have done this many times and been absolutely astounded at how quickly long-standing patterns in my life have completely changed! Know that when you perform this exercise, it is already done. Space and time do not exist in the moment of the cosmic blink.

Every time you read a story in this book about how someone's life changed by applying these principles, you can apply the same principles in your own way, to every aspect of your life. Do not be concerned about what you are doing because this does not involve doing anything. You are learning a new skill set that teaches you how to access and to evolve into your own unique appreciation and understanding of the secrets of the Matrix.

Here is how one Master Practitioner demonstrates his process of manifestation:

MANIFESTING COLESLAW

My third seminar was profoundly, earth-shatteringly different in ways incomprehensible to my conscious mind. It became exceedingly clear that I could connect with any person, place, or thing regardless of where I was (first point), or it was (second point). I realized that I did not have to go to the second point because I was already there; I am it.

I merely shift my awareness of me as me to my awareness of me as I'm there. My perception expands beyond space and time, and I enter into a state of being where existing in a magically whimsical universe becomes my favorite flavor of reality. The following is what showed up at lunch on one of those perfectly ordinary Matrix days.

I was being seated in the restaurant when a group of attendees who had already ordered invited me to join them. I read the menu and decided on a burger and coleslaw. Upon placing my order, I was informed that the restaurant had run out of coleslaw. Sorry, but that just felt wrong to me, so I put my hand up to see if I could find some anywhere in the restaurant. Instantly I found some; I felt it. I could see it in my hand. I turned to the waiter, held out my hand, and said, "Look. It's right here in my hand. You do have some. Could you please get me a small bowl?" He grumbled and left for the kitchen.

A few moments later he returned to tell me that the chef had looked everywhere and was absolutely certain that there was no coleslaw. At this point the kind folks who had so graciously invited me to join them began to lose patience with my insistence on having coleslaw. I assured them that if I could find it, feel it, and see it, it did exist and no one was going to convince me otherwise. That was purely and simply my new, improved reality.

The next moment I heard myself telling the waiter to "enter the kitchen and turn left. There will be a tall cabinet standing next to a shorter one. Inside the short one, behind the white tub" is where he would find my coleslaw. He grunted and headed toward the kitchen again.

Three minutes later the waiter popped out of the kitchen with a huge tray of coleslaw for me and all the others who had been given potato salad instead. He was bewildered and mumbled repeatedly that the coleslaw was right where I had said it was. I knew without knowing—without ever going into the kitchen—and I trusted what was showing up.

—WM, *Matrix Energetics Certified Master Practitioner*

BUSTING LOOSE FROM THE REALITY GAME

From one perspective, we create our problems. Utilizing the *All Power* of universal intent, we create the appearance of our problems so that we can have an interesting life. Robert Scheinfeld, author of *Busting Loose from the Money Game*, describes in his book the process wherein we create the holograms or virtual picture shows of our life's experiences. He believes that in a very real sense we are all the actors, the screenplay writer, and the director in the science fiction-romantic-comedy-action-drama we call our life.

Everything we observe and experience is a holographic projection played out on the cinema screen of our earthly existence. As the director of our drama, we can choose at any time to rewrite the screenplay and even recast or recreate the scenes of our earlier experiences so that they play out differently. This idea reminds me of an old song from Jethro Tull's concept album *Passion Play*. There is a verse in that song that says, "We've got you taped, you're in the play. How does it feel to be in the play? How does it feel to play the play? How does it feel to be the play?"

At the level of mind where the universal hologram is created and sustained, there is no such thing as poverty, disease, or suffering of any kind. These are just fluctuations and patterns of energy, which carry no negative context. So if what you have created no longer amuses you, then you can re-create, recast the players, and rewrite your experiences in a new way. Your life is like an episode of reality television and it is carefully scripted.

Scheinfeld says that everything you experience and see around you is just you following the script, which your Universal Self has given you to play with for your own growth and entertainment. When you start to realize this and you begin to reclaim your power, that energy comes back to you requalified and ready to be used for your next venture or reality simulation.

Here is an example of magical manifestation by one of my students, a masterful wizard:

We are publishers, but our medium for publishing is the Internet rather than print media. We have a variety of websites and earn income through monetizing them with affiliate links and Google Adsense, etc. Anyway, this is an experience I had prior to reading the book Busting Loose from the Money Game. *It led me to understand more deeply how to live in what is referred to as Phase-2 living, so that when I did read* Busting Loose from the Money Game, *I could embrace this natural state to greater degrees of conscious awareness.*

Dr. Bartlett's publicist contacted our company about reviewing his book Matrix Energetics: The Art of Science and Transformation. *Upon reading the description from his publicist, I had a hunch that Dr. Bartlett had a deep wisdom to share and that this was not just a rehashed technique or healing modality that was already available in some other form. So I personally requested a review copy. Days went by, and then one day, I went to the mailbox and found a package from the publicist. I felt a shiver when I retrieved the envelope from the mailbox . . . you know, like when you get that certain tap on your shoulder to pay attention. I went inside the house, ripped open the envelope, and stared at the cover. Pouncing on the couch, I began reading. I thought to myself . . . yup, this feels different, exciting, and promising.*

My partner noticed I was diverting my attention from my normal schedule and inquired what I was up to. All I could tell him was I had a feeling that I had found something that was going to bring together many things I had always known. Then I went back to reading.

I took the rest of the day and read the book from cover to cover (much like I did with Busting Loose from the Money Game, *which Dr. Bartlett had mentioned to the participants at one of his workshops that I later attended).*

As I began playing with some of the ideas in Matrix Energetics, I noticed a real excitement brewing in my being. We had been struggling with our business due to new algorithms with Google that were adversely affecting our website traffic, and our income had also dipped to an all-time low. The timing was perfect. What I was reading was giving me a renewed sense of hope. I wasn't sure how, but I felt I was being given a key—not to worry, but change my conscious perception.

Later that evening, an image of a green spiral grid floated across my internal visual screen. Hmmm, what to do with this? How might this be useful? Then I got the idea of playing with it in ways that Dr. Bartlett described in the book—incorporating a quantum two-point technique. So I created a place on this green spiraling grid that represented our websites and another point that represented Google. I stood in my office like a mad alchemist, raising my hands in the air, connecting dots, seeing little spiders jump out from point to point on the grid, eagerly moving about. It became a mesmerizing dance, and if someone had walked into the room and looked at me they, might have thought I was conducting an invisible orchestra.

This whole procedure took about three or four minutes. It was so fun. What the heck; no harm in playing, right? Well, here is where this story gets very interesting. Around twelve hours later, I sat down at my computer to work and glanced at our Web traffic. Something just didn't look right. There had to be some kind of error. How could that be? Our traffic had spiked and our income had also taken a much-needed turn for the better. The numbers looked like magic. (And they have remained on track ever since that day.)

I ran into my life and business partner's office and told him that life was never going to be the same! We scheduled a trip to meet Dr. Bartlett at one of his seminars. And we have been practicing Matrix Energetics ever since. The nature of Phase-2 living and the transformation art of Matrix Energetics seem like a marriage made in heaven. It seems to me that everything is connected; all things are always available—it just depends how and when you choose to "rotate your conscious perception" to notice what is always there, or what you imagine to be there, so that it can be there.

I hope that anyone who reads this story gets a sense on how future events can affect the present and past and how everything is connected. I believe that working with the concept of grids and templates enlivened with imagination, intention, and intuition might be an effective tool. When we play, we invoke magic because our mind is not interfering, fretting, and measuring what isn't.

—SB

When you begin to say consistently and without judgment to all aspects of your existence, "I love you. I release you from limiting patterns and reclaim you as pure universal source," then the energy you have invested begins to flow back to you. That reminds me of the final scenes of the movie *The Hitchhiker's Guide to the Galaxy*: After the earth has been destroyed, it is re-created and put back together just like it was before.

The more you intensify and turn up the rheostat of your awareness potential, the more access you are granted to magical realms. Different natural laws function here that allow you to perform feats that to the uninformed might appear to be miraculous in nature. The true nature of miracles is entirely consistent with natural law. It is only natural that universal consciousness is unlimited in potential and scope.

Within the enlightened potential state of universal being, the concept of limitation is unnatural. Tom Bearden states that the idea that there are laws that govern nature is just plain stupid. Bearden says, "You can do anything at all; it might just take you a lifetime to figure out how to do some things."[3] ***There are no limitations beyond what you can and cannot conceive of doing. Why wait a lifetime? Begin now!***

MATRIX GLOSSARY

Abrams, Albert: Born in San Francisco in 1863, Abrams founded radionics, a method for diagnosing and treating pathologies in the body using vibrational frequencies. A graduate of the University of California, he wrote several medical textbooks and eventually gained recognition as a specialist in diseases of the nervous system. In the course of his ongoing research, Abrams discovered that diseases could be measured in terms of energy, and he devised the Oscilloclast, an instrument with calibrated dials that enabled him to identify and measure disease reactions and intensities in blood samples.[1] *See also* path-Oscilloclast; radionics.

aether physics: Also spelled ether or æther, aether (αἰθήρ) was the poetic personification in Greek mythology of the clear upper air breathed by the Olympians. In ancient and medieval sciences, aether was a classical concept sometimes referred to as the fifth element in a number of theories, including alchemy and natural philosophy. In physics, it is a theoretical, universal substance predominantly believed during the nineteenth century to act as the medium for transmission of electromagnetic waves, such as light and x-rays, much as sound waves are transmitted by an elastic media like air. The aether was assumed to be weightless, transparent, frictionless, and undetectable chemically or physically—literally permeating all matter and space. The theory met with increasing theoretical opposition in 1881 by the Michelson-Morley experiment, which was designed specifically to detect the motion of the Earth through the aether but demonstrated no such effect.[2]

antiparticle/anti-wave: The time-reversal (phase-conjugation) of a reference wave, or, in other words, the anti-wave of the reference wave.[3]

archetypes: In Jungian psychology, any of several innate ideas or patterns in the psyche, expressed in dreams, art, and other subconscious processes as certain basic symbols or images. Archetypes act as the core focus around which the experiences and contents of our psyche are formed. "Everything in nature can be described in terms of geometry. From the dance of atoms to the revolutions of the planets, every type of growth and motion is governed by the same set of laws. These laws are portrayed through the symbolism of geometric shapes and forms. This was the domain of dreams, the unexpressed or unarticulated—images that, by their very design, were intended to engage the unconscious. The use of archetypal imagery provides a bridge between the realms of mind, the imagination, and the domain of physical manifestation."[4]

artificially generated quantum potential: A scalar potential is composed of, or partially contains, an artificially assembled bidirectional wave set. See E. T. Whittaker for proof that a "scalar potential" is actually a harmonic set of hidden bidirectional, longitudinal electromagnetic phase-conjugate wave pairs. Each wave pair consists of a wave and its anti-wave (true time-reversed replica wave). If the external observer could see the *detected* (effect) waves in a hidden wave pair, he or she would see the "wave" going in one direction and the anti-wave passing precisely through it in the other direction. However, prior to detection, the phase conjugate wave exists entirely in the complex plane and hence in the time domain.[5]

Bearden, Thomas: PhD, nuclear engineer, retired lieutenant colonel (U.S. Army), CEO, director of the Association of Distinguished American Scientists, and fellow emeritus of the Alpha Foundation's Institute for Advanced Study. Bearden is a theoretical conceptualist active in the study of scalar electromagnetics, advanced electrodynamics, unified field theory, KGB energetics weapons and phenomena, free energy systems, electromagnetic healing via the unified field action of extended Sachs-Evans electrodynamics, and human development. Particularly known for his work establishing a theory of overunity electrical power systems, scalar electromagnetic weapons, energetics weapons, and the use of time-as-energy in both power systems and the mind-body interaction.[6]

bioplasmic field: Dr. Victor Inyushin at Kazakh University in Russia suggests the existence of a so-called bioplasmic energy field composed of ions, free protons, and free electrons. His observations showed that the bioplasmic particles are constantly renewed by chemical processes in the cells and are in constant motion. There appears to be a balance of positive and negative particles within the bioplasma that is relatively stable. In spite of the normal stability of the bioplasma, Inyushin has found that a significant amount of this energy is radiated into space. Clouds of bioplasmic particles that have broken away from the organism can be measured moving through the air.[7]

Braun, Wernher von: One of the preeminent rocket developers and champions of space exploration from the 1930s to the 1970s.[8]

Broglie, Louis de: French physicist and mathematician who was chiefly devoted to the study of the various extensions of wave mechanics: Dirac's electron theory, the new theory of light, the general theory of spin particles, applications of wave mechanics to nuclear physics, etc. He published numerous notes and several papers on this subject and is the author of more than twenty-five books on the fields of his particular interests. In 1929 the Swedish Academy of Sciences conferred on him the Nobel Prize for Physics "for his discovery of the wave nature of electrons," published originally in 1924 as his doctoral thesis, *Recherches sur la théorie des quanta* (*Research on Quantum Theory*). In his later career, de Broglie worked to develop a causal explanation of wave mechanics in opposition to the purely probabilistic interpretation of Born, Bohr, and Heisenberg, which now dominates quantum mechanical theory. De Broglie also made major contributions to the fostering of international scientific cooperation.[9] *See also* Dirac sea.

calibrate: A way of noticing patterns of energy and information and assigning some form of real-time measurement to that experience. Measuring pre- and post-changes after a Two-Point is one very useful form of calibrating a measurement in Matrix Energetics.

carrier wave: A fundamental wave that is modulated by another wave or other waves and "carries" the other modulating waveform(s). Stripping the carrier in a demodulator will reveal the carried waveform(s).[10]

Casimir effect: Widely cited as evidence that underlying the universe is a sea of real zero-point energy. In 1947 physicist Hendrik Casimir had the opportunity to discuss ideas with Niels Bohr on a walk. According to Casimir, Bohr "mumbled something about zero-point energy" being relevant. This led Casimir to an analysis of zero-point energy effects in the related problem of forces between perfectly conducting parallel plates.[11] *See also* zero-point energy.

closed system: In the present approach, a system that in theory does not communicate with its environment and does not exchange energy or matter between system and environment. There is really no such thing as a truly closed system in the universe since every system is embedded in the active vacuum and is an open system in energy exchange with the vacuum.[12] *See also* open system.

coherence: A correlation between the phases of two or more waves so that interference effects may be produced between them, or a correlation between the phases of parts of a single wave.[13] *See also* decoherence.

collapse of the wave function: Also called a state-vector collapse or wave-packet reduction, a wave function collapse is one of two processes by which quantum systems evolve in time, according to the laws of quantum mechanics. The concept was originally introduced by Werner Heisenberg in his uncertainty paper and later postulated by John von Neumann as a dynamical process independent of the Schrödinger equation.[14]

complex conjugate number: Extensions of real numbers, the complex conjugate of any real number has the same real component accompanied by a Greek ι or i, which transforms the plane so that all points are reflected on the real axis—that is, points above and below the real axis are exchanged while points on the real axis remain the same (since the complex conjugate of a real number is itself).[15]

decoherence: A non-unitary process that describes a thermodynamically irreversible disturbance (change of state) of the environment by the system—rather than distortion of the system by its environment. This gives the *appearance* of wave function collapse when a system interacts with its environment, which prevents different elements in the quantum super-position of the system and environment's wave function from interfering with each other.[16] *See also* coherence.

Dirac, Paul: The importance of Dirac's work lies essentially in his famous wave equation, which injected special relativity into Schrödinger's equation. Taking into account the fact that, mathematically speaking, relativity theory and quantum theory are not only distinct from each other but also oppose each other, Dirac's work could be considered a fruitful reconciliation between the two theories.[17]

Dirac sea: A theoretical model of an infinite sea of negatively energized particles in a vacuum. Erwin Schrödinger is said to have been the first to note that solving the Dirac equation for the motion of the electron resulted in a necessary component that could be interpreted as random, speed-of-light fluctuations of a point-like particle. He dubbed this motion *zitterbewegung,*

German for "jitter motion." The positron, the antimatter counterpart of the electron, was originally conceived of as a hole in the Dirac sea, well before its experimental discovery in 1932. Dirac, Einstein, and others recognized that it is related to the aether.[18] *See also* electron; Schrödinger, Erwin.

double-slit experiment: The experiment that demonstrates the concept in quantum mechanics that energy-carrying waves can also behave like particles and that particles can also display a wave aspect—just not at the same time. A light source illuminates a thin plate with two parallel slits cut in it, and the light passing through the slits strikes a screen behind them. The wave nature of light causes the light waves passing through the slits to interfere with one another, creating a pattern of bright and dark bands on the screen. The screen, however, absorbs the light as discrete particles called photons.

drop down: Most people in our society could be said to be centered in their heads, entertaining the mistaken belief that this is where consciousness arises. In the prevailing Western scientific model, the human brain is thought to generate the process of consciousness, and the brain and mind are considered to be inseparable, if not one and the same. However, there is a different form of consciousness that could be called heart-centered awareness, or what Daniel Goleman refers to as "emotional intelligence." This form of awareness is based on letting go of the distracting interplay of mental chatter and dropping down into the theta state (7–4 cycles per second). You can learn, as military remote viewers do, to consciously sustain this state and operate from within it. When you do this, you begin to consciously access the realm of the right brain, or the subconscious awareness. The state of dropping down is characterized by a particular awareness in which you notice but do not analyze or judge.

electromagnetic energy: From a vacuum and quantum mechanical viewpoint, a deterministic or coherent structuring, either dynamic or static, existing in the virtual-photon or charged-particle flux. From a space-time viewpoint, a curvature of space-time or set of such curvatures.[19]

electron: A stable elementary particle in all atoms having—in "forward time"—a negative charge of 1.602 x 10^{19} coulombs, a spin 1/2, and a mass of 9.11 x 10^{-31} kilograms. If time is reversed, the charge (but not the mass) of the electron is reversed and it becomes a positron. The "electron" may also exist as a negative energy, negative charge, or negative mass energy in the vacuum itself. In this state, the negative energy is the source of negative energy fields and negative energy potentials. The hole in the Dirac sea created by this positron can be manipulated in "anti-circuits" to directly engineer local antigravity, quite strongly and practically.[20] *See also* Dirac sea.

electron cloud: *See* probability cloud.

energetic rapport: Describes the process of being on the same wavelength as the person, place, thing, pattern, or expression of energy with whom you are relating. Techniques that are beneficial in building rapport include matching your body language (e.g., posture and gesture), maintaining eye contact, and matching breathing rhythm. Some of these techniques are explored in neuro-linguistic programming. Energetic rapport is not accomplished through a mental process but by dropping down into heart-centered awareness and matching up with the feelings and imagery of the object, energy, or pattern you are attempting to synchronize with. This form of rapport is facilitated by a process of noticing whatever it is you notice without judging or analyzing and then allowing yourself to be guided step by step by whatever occurs in the next moment. *See also* drop down.

entropy: In thermodynamics, a measure of disorder in a system that quantifies the amount of energy unavailable for useful work in a system undergoing change, such as the expansion of a gas in a vacuum or the heat transfer from a hot body into a cold body. These changes cause an increase in entropy for the system under consideration, but energy is not transferred into or out of the system. In other words, as entropy rises, the available energy of the system decreases. *See also* negentropy.

ether: *See* aether physics.

Feynman, Richard: An American physicist who was a key figure in "the development of the theory of quantum electrodynamics, laying the foundation for all other quantum field theories. His approach combined quantum mechanics and relativity theory and exploited a method using diagrams of particle interactions to greatly simplify calculations. For this work, he shared with American physicist Julian Schwinger and Japanese physicist Sin-Itiro Tomonaga the 1965 Nobel Prize for physics."[21]

focused intent: Can be defined as the creative act of using the many and varied parts of your total conscious experience to define a set of new experiences, realities, or outcomes in your current experience. It is helpful to focus your imagination in order to create a new sensation. This will initiate a flow of subtle energy to directly or indirectly influence or manifest the desired events and effects, thus to create or focus with feeling. The reason for focusing your intent is to convince yourself that you can "move" into that new reality. Since every application of our intention is an act of creation, focused intent ultimately teaches us how to create efficiently and effectively. This in turn ultimately manifests in some type or types of events in our sensory world. The object of your focused intent is like the blueprint for manifestation. However, once the blueprint is conceived, you must then let go of it to the universe in order for the form to be given life. This is the alchemical process for manifestation.[22]

Garcia, Hector: Certified Master Practitioner of Matrix Energetics and founder of the Garcia Chiropractic Holistic Center in San Diego, California, where he uses his unique intuitive gift as a medical intuitive to detect physical, mental, emotional, psychological, psychic, and spiritual imbalances that show up in different energy systems of the body. Whether on a cellular dimension or on the quantum level, Garcia is able to pinpoint the root cause of the problem, facilitate the body to release it energetically, and correct the issues. As a trained and skillful chiropractic diagnostician, he uses several techniques, such as Contact Reflex Analysis, a nutritional balance assessment; Neuro Emotional Technique, an emotional release therapy; Allergy Elimination Technique, a method that removes sensitivities to allergens; and the Yuen Method of Chinese energetic medicine (of which he is an instructor), an energetic healing technique that can detect any imbalances or deficiencies affecting the body and its capacity to heal itself.[23]

general theory of relativity: Einstein's theory of gravity, in which the gravitational force is represented by a curvature in space-time and in which space-time is an active entity. We may look at all forces as being due to curvatures of space-time interacting with mass.[24] *See also* special relativity model.

graviton: In the quantum theory of gravitation, the graviton is the quantum of the gravitational field. It is massless. In the new theory, we may take the graviton as a coupled scalar and longitudinal photon pair.[25]

Hamilton, William Rowan: An Irish physicist, astronomer, and mathematician who made important contributions to classical mechanics, optics, and algebra. His studies of mechanical and optical systems led him to discover far-reaching mathematical concepts and techniques.[26]

heart field: "The heart is the most powerful generator of electromagnetic energy in the human body, producing the largest rhythmic electromagnetic field of any of the body's organs. The heart's electrical field is about sixty times greater in amplitude than the electrical activity generated by the brain. This field, measured in the form of an electrocardiogram (ECG), can be detected anywhere on the surface of the body. Furthermore, the magnetic field produced by the heart is more than five thousand times greater in strength than the field generated by the brain, and can be detected a number of feet away from the body, in all directions, using SQUID-based magnetometers (superconducting quantum interference devices that measure extremely small magnetic fields)."[27] Matrix Energetics teaches that through the heart field you have access to the zero-point energy field, the unlimited potential energy of the universe. *See also* right brain; zero-point energy field.

Heaviside, Oliver: Noted English self-taught physicist and brilliant electrodynamicist who played a role in discarding Maxwell's quaternions and also in forming vector mathematics and formulating the vector reduction of Maxwell's theory from twenty quaternion equations in some twenty unknowns to the present four vector equations.[28]

Heisenberg, Werner: German physicist whose cat thought experiment theory and its applications resulted especially in the discovery of allotropic forms of hydrogen. Heisenberg was awarded the Nobel Prize for physics in 1932. His theory is based only on what can be observed—that is, on the radiation emitted by the atom. We cannot always assign to an electron a position in space at a given time, nor follow it in its orbit, and we cannot assume that the planetary orbits postulated by Niels Bohr actually exist. Mechanical quantities, such as position or velocity, should be represented not by ordinary numbers but by abstract mathematical structures called "matrices," and he formulated his new theory in terms of matrix equations. Later, Heisenberg stated his famous principle of uncertainty, that the determination of the position and momentum of a mobile particle necessarily contains errors the product of which cannot be less than the quantum constant h and that, although these errors are negligible on the human scale, they cannot be ignored in studies of the atom.[29]

Herbert, Nick: A PhD in experimental physics, Herbert was senior scientist at Memorex and other Bay Area hardware companies specializing in magnetic, electrostatic, optical, and thermal methods of information processing and storage. He is the author of *Quantum Reality: Beyond the New Physics, Faster Than Light* (published in Japan under the title *Time Machine Construction Manual*) and *Elemental Mind: Human Consciousness and the New Physics*. Herbert devised the shortest proof of Bell's interconnectedness theorem to date. He has written on faster-than-light and quantum theory for such journals as the *American Journal of Physics* and *New Scientist* and is a columnist for *Mondo 2000*.[30]

hologram: "One of the things that makes holography possible is a phenomenon known as interference. Interference is the crisscrossing pattern that occurs when two or more waves, such as waves of water, ripple through each other. Once they've collided, each wave contains information, in the form of energy coding, about the other, including all the other information it contains. Interference patterns amount to a constant accumulation of information, and waves have a virtually infinite capacity for storage. When a complex series of interference patterns interact, they form a hybrid of highly structured information, which is the building block of what we perceive as our reality constructs. If you drop a pebble into a pond, it will produce a series of concentric waves that expands outward. If you drop two pebbles into a pond, you will get two sets of waves that expand and pass through one another. The complex arrangement of crests and troughs that results from such collisions is known as an interference pattern." The snapshot of the pond's surface is similar to a holographic image in that it consists of a set of interference patterns produced by the combining of the various wave fronts. Neurosurgeon Karl Pribram realized that if the holographic brain model was taken to its logical conclusions, it opened the door on the possibility that objective reality—the world of coffee cups, mountain vistas, elm trees, and table lamps—might not even exist, or at least not exist in the way we believe it does.[31]

imaginary unit (number): In mathematics, physics, and engineering, the imaginary unit is the square root of negative one. Denoted by the letter i or the Latin $\dot{j}$ or the Greek ι, the unit allows the real number system to be extended to the complex number system.

inertial frames: If two systems are moving uniformly in relation to each other, one cannot determine anything about their motion except that it is relative. Each of the two frames is said to be "rotated" with respect to the other, but not accelerating. The velocity of light in space (the vacuum) is constant and independent of the velocity of its source and the velocity of an observer. All the laws of physics are the same in all inertial frames of reference.[32]

innocent perception: A cultivated practice of allowing perception to be the gathering arm of consciousness so it may stretch beyond our former reach. The ongoing search for consciousness directs all our perceptions. We actually *see* with our total consciousness, not our eyes, which are only data-gathering instruments.[33]

least-action pathway: Least-action paths are the routes we observe in everyday life. They are also literally the pathways for neural information in our brains and nervous systems. By becoming conscious of the world around us, we create these pathways. They become habits necessary for our survival. At a quantum level, the action of the path can change depending upon the observer of that path. Each observation creates a least-action connection with the previous observation. By choosing to observe reality along a particular path, what was unobserved then becomes a greater-action path, even if it would have been a least-action path if it had been observed. In other words, by choosing to observe a particular path over others, a least action is created. Consciousness creates, from all the paths, the one that takes the least action.[34]

left brain: In each second, consciousness reveals to us a tiny fraction of the eleven million bits of information our senses pass on to our brains. While the right brain functions along with the left and can process billions of bits of data per second, the left brain is a serial processor that can only handle seven, plus or minus two, bits of information per second. In this way, the left brain, the namer and classifier, functions as the analytical sensor, determining what sensory data is allowed to be perceived consciously and what gets pushed below the threshold of the conscious awareness. This is useful to navigate in the so-called real world. However, because the left brain censors what is allowed and what is disavowed in our sensory awareness, massive amounts of data are continually deleted and considered negligible by the left brain's innate programming. As soon as you name something, you have defined it. In defining it, you have dictated how it shows up in your reality. This is why the Norwegian scientist Tor Nørretranders says, "Trust your hunches and intuitions—they are closer to reality than your perceived reality, as they are based on far more information."[35] *See also* letting go; right brain.

letting go: The left brain (analytical mind) has been known as the "monkey" mind. The way to trap a monkey in the jungle is to set out a box with a piece of fruit in it with a hole in the box that is only large enough for the monkey's empty hand to enter. When the monkey grabs the fruit in the box, it cannot get its hand out of the box without first letting go of the fruit. But a majority of monkeys never let go, which makes them easy catches for the hunters who lay such traps. This analogy characterizes the behavior of the left-brained individual. Once you decide what the rules are and how they should show up, you can spend your whole life holding on to the fruit of your ideas and never let go in order to experience a larger and more expansive sense of reality. In Matrix Energetics, students are taught how to drop their awareness out of their head, out of a concerned and judgmental mind-set, and into the field of the heart. The field of the heart links naturally with the domain of the right brain and its emotional intelligence. It is from this latter domain of consciousness, the field of dreams, as it were, that the greatest ideas and creative concepts have been conceptualized and realized. This is what Matrix Energetics means by "letting go." By letting go of the need to understand your experiences, particularly if they fall outside your perceptual box, you can begin to trust the feeling-oriented state of heart-based awareness.

many-worlds theory: The fundamental idea of the many-worlds theory was proposed by physicist Hugh Everett in 1957: that there are myriads of worlds in the universe in addition to the world we are aware of. In particular, every time a quantum experiment with different outcomes with non-zero probability is performed, all outcomes are obtained, each in a different world, even if we are aware only of the world with the outcome we have seen. In Matrix Energetics, quantum experiments take place everywhere and very often, not just in physics laboratories; even the irregular blinking of an old fluorescent bulb is a quantum experiment.[36] *See also* parallel universe

Matrix Energetics: The art and science of transformation.

Maxwell, James Clerk: Scottish mathematician and theoretical physicist whose most significant achievement was the development of the classical electromagnetic theory, synthesizing all previous unrelated observations, experiments, and equations of electricity, magnetism, and optics into a consistent theory. His set of famous equations demonstrated that electricity, magnetism, and light are all manifestations of the electromagnetic field. From that moment on, all other classical laws or equations of these disciplines became simplified cases of Maxwell's equations. Ivan Tolstoy, in his biography of Maxwell, wrote, "Maxwell's importance in the history of scientific thought is comparable to Einstein's (whom he inspired) and to Newton's (whose influence he curtailed)."[37]

Maxwell's electrodynamics (electromagnetic theory): Simply put, Maxwell's theory of electrodynamics consists of his equations. His fundamental theory consisted of some twenty quaternion equations in twenty unknowns, appearing in his 1865 paper. After Maxwell's death and some curtailment by Maxwell himself, Oliver Heaviside modified and sharply curtailed these equations into the familiar four equations of today, as did Willard Gibbs and Heinrich Hertz. Hendrik Lorentz further curtailed the Maxwell-Heaviside equations by symmetrically regauging them. This simplified their mathematical solution but it also inadvertently and arbitrarily discarded all open Maxwell systems far from thermodynamic equilibrium with their active environment (such as the modern active vacuum).[38]

Maxwell's unified field theory: The term is usually applied with respect to any physical or electromagnetic system whose electrodynamic operations obey Maxwell's electrodynamics model after the Lorentz regauging of the Maxwell-Heaviside equations. This subset is now unfortunately and inappropriately referred to as "Maxwell's equations." As a result, many scientists and most engineers today no longer understand Maxwell's theory. When embedded in a higher-topology algebra, the theory permits a vast richness of additional electromagnetic systems and behaviors, including full unified field theory operations. In standard gauge symmetry electrodynamics, all these higher-symmetry functions and systems are excluded a priori. From Thomas Bearden's view, this is particularly sad since these arbitrary reductions of Maxwell's theory exclude all electromagnetic systems far from equilibrium in their exchange with the active vacuum. Hence, almost all scientists and engineers believe it is against the very laws of nature to propose an electrical power system that produces more energy output—and more work in the load—than the energy input made by the operator himself.[39] *See also* Heaviside, Oliver.

McMoneagle, Joe: American remote viewer, No. 001 (372), who has provided paranormal support to the Central Intelligence Agency, Defense Intelligence Agency, National Security Agency, Drug Enforcement Agency, Secret Service, Federal Bureau of Investigation, United States Customs, National Security Council, and Department of Defense. He was one of the original officers recruited for the top-secret army program now known as the Stargate Project.[40]

module: A module is a self-contained component of a system that has a well-defined interface to other components of the system. In essence, Matrix Energetics developed an interdimensional technology that holds a space from the zero-point field and creates a conceptual bridge point from the imaginary to the real. This describes, by the way, a complex conjugate number in quantum physics, where minus one represents the imaginary quality, and then there is the real number, or quality, as well. When you multiply the two together, the imaginary component cancels out, and you end up with a real vector or coordinate in space-time. This is exactly how this energetic technology works; it is a spiritual technology that it is engineerable and reproducible. Think of a module as creative, interactive spiritual software designed to perform or address or correct a given task, condition, or activity. A module in this sense could be said to perform the function of what Dr. William Tiller calls an intention-imprinted device. *See also* complex conjugate number; imaginary unit; Tiller, William.

morphic field: A field within and around a morphic unit, which organizes the field's characteristic structure and pattern of activity at all levels of complexity. The term *morphic field* includes morphogenetic, behavioral, social, cultural, and mental fields. Morphic fields are shaped and stabilized by morphic resonance from previous similar morphic units, which were under the influence of fields of the same kind. They consequently contain a kind of cumulative memory and tend to become increasingly habitual.[41]

morphic resonance: Coined by Rupert Sheldrake in his 1981 book *A New Science of Life*, resonance on a morphic level is "the influence of previous structures of activity on subsequent similar structures of activity organized by morphic fields. Through morphic resonance, formative causal influences pass through or across both space and time, and these influences are assumed not to fall off with distance in space or time, but to come only from the past. The greater the degree of similarity, the greater the influence of morphic resonance." The expression of resonance, therefore, refers to what Sheldrake thinks is "the basis of memory in nature . . . the idea of mysterious telepathy-type interconnections between organisms and of collective memories within species."[42]

morphic unit: A unit of form or organization, such as an atom, molecule, crystal, cell, plant, animal, pattern of instinctive behavior, social group, element of culture, ecosystem, planet, planetary system, or galaxy. Morphic units are organized in nested hierarchies of units within units: a crystal, for example, contains molecules, which contain atoms, which contain electrons and nuclei, which contain nuclear particles, which contain quarks.[43]

motionless electromagnetic generator (MEG): The transformer-like electrical power generator, invented by Thomas Bearden, James Kenny, James Hayes, Kenneth Moore, and Stephen Patrick, that powers the transformer core with a permanent magnet but separates curled magnetic vector potential from uncurled magnetic vector potential so that the magnetic field flux is retained in the core while the potential is replenished outside the core and adjacent to it.[44]

negative time: In quantum mechanics, each virtual photon is continually turning into an electron/positron pair, and vice versa. Paul Dirac proposed that a positron is an electron traveling backward in time. Further, pair production produces time-smeared particles—an electron and a positron. Pair production, then, actually produces two electrons: one coupled to (smeared in) a positive piece of time and one coupled to (smeared in) a negative piece of time. Thus, in the vacuum, two discrete streams of time, one positive and one negative, are continually created in conjunction with pair production and destroyed by pair annihilation. Further, integration of tiny virtual pieces of time (from virtual photons) to form "passage of time" macroscopically is directly associated with the charge (the absorption and emission of virtual particles) of an observable particle. This is what is meant by an object "existing" (persisting). Its continual virtual-photon interactions are integrated by its mass (timeless) part in comparatively larger jumps through time. The absorption of the photon connects a positive piece of time to the mass of the particle, converting it to mass time. The subsequent emission of an observable photon, "tears of the little time-tail," so to speak, leaves behind a totally spatial mass entity, having no large connection to "the flow of time."[45]

negentropy (negative entropy): In a sense, the reversal of disorder, or reversal of entropy. In a negentropic system, energy would proceed from a state of disorder to increasing order. In a biological system, such as a human body, this would describe the principle of a self-organizing system.[46] *See also* entropy.

Neumann, John von: Hungarian-American mathematician who developed the branch of mathematics known as game theory. In 1933 he joined the Institute for Advanced Study in Princeton, New Jersey, and later served as a consultant on the Los Alamos atomic-bomb project during World War II. In 1955 he became a member of the U.S. Atomic Energy Commission. Von Neumann is noted for his fundamental contributions to the theory of quantum mechanics, particularly the concept of "rings of operators" (now known as Neumann algebras), and for his pioneering work in applied mathematics, mainly in statistics and numerical analysis. He is also known for the design of high-speed electronic computers.[47]

Newtonian physics (Newton's laws of motion): From Sir Isaac Newton's *Philosophiæ Naturalis Principia Mathematica*, the three physical laws that form the basis for classical mechanics. Newton used them to explain and investigate the motion of physical objects and systems, including the motion of earthbound objects and planetary motion. *First law:* Often referred to as the law of inertia, a body will persist in a state of rest or uniform motion unless acted upon by an external unbalanced force. *Second law:* Observed from an inertial reference frame, force equals mass times acceleration. *Third law:* To every action there is an equal and opposite reaction.

not doing: To try to understand is a matter of *doing*. Clearly, it is easier to explain *doing*. It seems to be akin to objectivity; a rock is a rock because of doing—"because of all the things you know how to do to it." It is important that without doing, nothing is familiar. If nothing is familiar, everything is new and unknown and experienced for the first time—it is unconditioned. When you try "to figure it out, all you're really doing is trying to make the world familiar." This is doing and involves rational activity—specifically rational or intellectual formulation of experience. Acting without belief is *not doing*. Belief, again, orders experience in an attempt to make it meaningful. The technique of *not doing* is facilitated by displacement of normal doing by a different doing in a trick analogous to learning an alternative world description. In the same way, both worlds—the world we all know and the sorcerer's world—are unreal, but they are useful, even if not necessary models of reality.[48] In Matrix Energetics, it is taught that whatever you are doing, you are doing it with your perceptual model of how things get done. In other words, whenever you enter into the state of doing something, you are working off your previous assumptions about how the world works and how things are. When you suspend these judgments and merely perceive an act in the moment, you are entering into the realm of grace, which is characterized by the art of not doing. An old and perhaps trite corollary of this is to "let go and let God." Every act of doing is performed by the doer, the left-brain mentality. We believe that what we do makes a difference, yet how easily are our plans and our schemes undone by life's circumstances? The illusion of control over the energy patterns of the universe is just that. When we humbly acknowledge, even as Jesus did, that "I of my own self can do nothing," we are now seated in the throne of the heart's domain, having surrendered our human conceptions concerning what is possible or impossible. Now things are done through us, not by us. Ultimately, the act of not doing is a simple act of faith that there is a higher power that, given the opportunity, can and does act through us.

noticing: In Matrix Energetics, noticing is the art of paying attention to what shows up in the moment by forming the habit of asking the open-ended question, "What am I noticing now?" You begin to train the unconscious mind to allow more information in from the domain of right-brain awareness to consciously held awareness. By doing this consistently, you begin to train yourself to become aware of energy and information that support your newfound dictate to pay attention to new stimuli and patterns in your environment (both internal and external). A corollary to this is to notice what is different, not the same. When consistently done, this trains the brain to look for new patterns and behaviors and reinforces any beneficial changes.

observational frame of reference: A spatial, organized, measured lattice placed in "emptiness" (space, space-time). Normally refers to a three-dimensional spatial frame. All objects and points in the "universe" or spatial frame are considered to simultaneously coexist at separate, measured points in the frame. Differs from the vacuum in that, rigorously, vacuum has no existing definite lengths and no existing definite time intervals, as these appear only after measurement or detection and are relative to the observer and to the interactions ongoing as well as in the detection process itself. The "laboratory frame" is the static reference frame of the observer or measurement. A separate reference frame may be assumed to exist for any fixed or moving object, or centered on any point in another frame. When a type of frame is assumed, the entire class of physical interactions that can occur has been restricted to an assumed set or type. In other words, given the frame, the *conventional* physics has been assumed. One of the greatest restrictions of an assumed "frame" is to rule out the consideration (existence) of other higher dimensions. Note that in the new unified field theory approach, the other higher dimensions are always available and cannot be ruled out in general, but only in some special cases. Every curvature of space-time, and any internal additional curvature to that primary curvature, adds a new dimension. In our view, a space-time may be "flat" in overall curvature but consist of internally structured deterministic curvatures or "engines." In this view, normal inertial frames, for example, may still contain vacuum engines, which will not affect the normal bulk translation rules but may affect any or all the non-translation mechanisms, including the very laws of nature in many cases.[49]

observer effect: The changes that the act of observation will make on the phenomenon being observed. This is often the result of instruments that, by necessity, alter the state of what they measure in some manner. This effect can be observed in many domains of classical and quantum physics.

open system: A system that communicates with its environment and exchanges energy and/or matter with its environment. With the possible exception of a few theoretical or hypothetical systems, all systems in the universe are, in fact, open systems. An open system may be in equilibrium with its active environment so that it cannot accept, store, and utilize any excess energy from the environment.[50] *See also* closed system.

Oscilloclast: *See* path-Oscilloclast.

paired spin: The intrinsic angular momentum of a particle, such as an electron, proton, neutron, photon, or graviton, for example—even when at rest, as if it were a top spinning about an axis but had to spin 720 degrees before it turned "full circle." Spin is quantized, and always described, as a half or whole spin (- 1, - 1/2, 0, 1/2, 1, and so on). A spinning charged particle, such as an electron, thus demonstrates a magnetic moment, due to the circulation of charge in the spinning. In the nucleus of an atom, the spin of the nucleus is the resultant of the spins of the nucleons (particles comprising the nucleus). Spin of particles would appear to be more like an "implosion" to "explosion" circulation. In other words, the particle circulates in the time domain (complex plane) as well. It would appear that the spin of a particle is the basic feature that integrates the (disintegrated) flux energy of vacuum flux into observable charge. Apparently, all observable fields, matter, effects, and so forth depend upon this basic mechanism to zip together virtual entities and form observable phenomena.[51]

parallel processing: The ability of the brain to simultaneously, seamlessly process incoming stimuli for quick, decisive action.

parallel universe: Parallel universes are self-contained, whole universes, infinite in number, and exactly the same as the one next to them except for one change. Moving across the infinite number of them, of course, you can get to any change you might want. These universes are all related to ours; indeed, they branch off from ours, and our universe is branched off from others. Within these parallel universes, our wars have had different outcomes from the ones we know. Species that are extinct in our universe have evolved and adapted in others. In other universes, we humans may have become extinct. A specific group of parallel universes is called a multiverse.[52] *See also* many-worlds theory.

particle physics: The branch of physics using accelerators to study high-energy particle collisions that determine the properties of atomic nuclei and other elementary particles.[53]

path-Oscilloclast: The instrument that Dr. Albert Abrams invented for the treatment of disease. The word means "vibration breaker." After being attached to a power source, the device is connected to the patient's body. By means of its rheostat, various vibratory rates may be produced. If a patient is suffering from tuberculosis, the Oscilloclast is set to emit a vibratory rate into the body corresponding to the rate that the disease has already created in the system. The patient feels no sensation because the vibrations are smaller than what human sense may detect.[54]

phase-conjugate: In nonlinear optics, the novel nonlinear mixing of waves that generate an output wave—called the phase-conjugate replica or time-reversed replica—that precisely retraces the path previously taken by the input wave that stimulated the action.[55]

possibility waves: Although observation is evidently necessary to bring about the transition from possible to actual, the fundamental nature of observation in quantum theory remains somewhat mysterious. This problem of measurement derives from the fact that prior to observation, the quantum is described as being a nonlocal wave of probability spread throughout space, while after observation, only one of the possible values is actualized. Thus, observation involves a discontinuous collapse, also called a "projection," of the quantum wave function from a continuum of possibilities to a single actualized value. This projection, however, is an ad hoc element of the formalism, and is not a lawful transformation that is governed by Erwin Schrödinger's wave equation. There is no explanation for how, when, or where this mysterious projection happens. Moreover, when the projection takes place, the laws of quantum physics do not predict which of the possible values will be actualized in any given observation, thus violating classical determinism and introducing an element of acausality and spontaneity into the theory at a fundamental level.[56] *See also* observer effect.

probability cloud: A term coined by physicist Richard Feynman in his discussion on "exactly what is an electron?" The electron cloud is often referred to as an orbital because the space where an electron is likely to be found cannot be actually pinpointed. The model provides a simplified way of visualizing an electron as a solution of Erwin Schrödinger's paradoxical thought experiment that demonstrated the randomness of a cat's life or death in a concealed box. In the electron cloud analogy, the probability density, or the distribution of electrons, is described as a small cloud moving around the atomic or molecular nucleus, with the opacity of the cloud proportional to the probability density.[57] *See also* Feynman, Richard; Schrödinger, Erwin.

psychotronics: The science of mind-body-environment relationships, an interdisciplinary methodology concerned with the interactions of matter, energy, and consciousness.[58] *See also* Abrams, Albert; radionics.

quantum field theory: A quantum mechanical theory in which a physical field is considered a collection of particles and forces. Observable properties of an interacting system are expressed as finite quantities rather than vectors.[59]

quantum/information potential: A special potential added to the Schrödinger equation by David Bohm in his hidden-variable theory of quantum mechanics. The quantum potential is a multiple connected entity, so it "occupies" widely separated but connected points, events, or objects. It is also an extraordinary energy amplifier, since any energy input to one of the multiple connected points simultaneously and instantly appears in every other connected point, regardless of distance or location in the universe. In real life, the quantum potential also has a "coefficient of multiple connectivity" so that only a fraction of the energy input to one multiple connected point will appear in the other points of the multiple connection group. The quantum potential has been strongly weaponized by five nations, and quantum potential weapons are the dominant weapons on earth, being more powerful than nuclear weapons. In theory, the quantum potential together with engines and anti-engines could be used to treat and cure a given disease in all persons on earth simultaneously. Sadly, engines have been developed to generate diseases in a targeted populace, rather than heal the populace. Russia and Brazil have had the quantum potential weapon for some time, as have two nations friendly to the United States. In 2001 China also deployed the quantum potential weapon.[60]

quaternion algebra: The algebra of quaternions and their mathematical operations. Quaternion algebra is of higher topology than either vector or tensor algebra. James Maxwell's original 1865 equations are some twenty equations in twenty unknowns, in quaternion and quaternion-like algebra. Oliver Heaviside and others reduced the algebra to vector algebra and some four equations. Maxwell himself was rewriting the equations in his treatise to purge the quaternions because of the controversy over the difficult quaternion form. The present vector equations taught in university as "Maxwell's equations" are, in fact, Heaviside's truncation of the Maxwell theory. Going to tensor algebra electrodynamics does not recover the full range of Maxwell's original 1865 quaternion electromagnetic theory.[61] *See also* Heaviside, Oliver; Maxwell, James Clark.

quaternion expression: Composed of the sum of four terms, one of which is real and three of which contain imaginary units, where the terms can be written as the sum of a scalar and a three-dimensional vector.[62]

radionics: A method of diagnosis and treatment at a distance that utilizes specially designed instruments with which practitioners can determine the underlying causes of diseases within living systems, be they human, animal, plant, or the soil itself. While radionics is mainly used to diagnose and treat human ailments, it has also been used in agriculture to increase yields, control pests, and enhance the health of livestock. *See also* psychotronics.

reference frame: Your set of rules for noticing what you notice; your construct for how things show up. The frame is the context through which you consider a situation—the frame of reference. The frame is incredibly powerful. Frames are generally unconscious filters for a situation. Because they're often unspoken and unrecognized, they can bypass our critical faculties and go straight into the unconscious. Every decision we make, the very way we see things as part of our decision-making process, is part of our unconscious frame of reference. Over time, we build up a set of rules around how to view and deal with a given type of situation. In order to deal with a given situation or to conceptualize a particular problem, we apply this preexisting software or these reference frames to see how any given situation or occurrence fits our perceptual model. This is largely done at an unconscious level. Accessing altered states of consciousness and then looking at familiar situations or patterns through a new perceptual lens can modify old reference frames or create new ones. Don Juan, Carlos Castaneda's teacher, might refer to this as the difference between "looking"—perceiving through the old frame of reference—and "seeing"—observing free from the usually distorted lens of our perceptual bias or habitual frame of reference.

renormalization: The procedure in quantum field theory by which divergent parts of a calculation, leading to nonsensical infinite results, are absorbed by redefinition into a few measurable quantities, so yielding finite answers.[63]

right brain: The right brain is a parallel processor that functions in a nonverbal manner and excels in visual, spatial, perceptual, and intuitive information, excelling with complexity, ambiguity, and paradox. Much faster than the left brain, the right brain processes information quickly in a nonlinear and nonsequential way, looking at the whole picture and billions of bits of information per second, and then determining the spatial relationships of all the parts as they relate to the whole. This component of the brain is not concerned with things falling into patterns because of prescribed rules. In Matrix Energetics, the right brain is conceptually linked to the person's heart field. This partnership of right hemisphere and heart, when properly developed, gives you the ability to access altered states of consciousness and unique perceptual frames of reference. *See also* heart field; left brain.

rule: A statement that tells observers how to render particular elements or aspects of a specific pattern of energy or experience. Your individual rule, or rule set, determines how the box of your virtual reality is constructed. Some possibly useful elements to include in your rule set are (1) a neutral point of view, (2) objectively verifiable means, and (3) a flexibility of viewpoint and its application. Bruce Lee elucidated three simple rules that governed the philosophy of the martial art called Jeet Kune Do, "the way of the intercepting (open) fist." His rule set had three principles: absorb what is useful, discard the classical mess (question habitual ways of thinking and acting that can lead to mechanical responses rather than creative spontaneity), and no way as way (which in Matrix Energetics means to stay neutral and flexible, notice what is different not the same, and act from your heart, not from the dictates of your head).[64]

Saint Germain: The Ascended Master Saint Germain teaches that the highest alchemy is the transformation of one's human consciousness into the divinity of the Higher Self. He stands ready to assist all souls in this endeavor. He has also said he would release the technology of the Aquarian age when the nations shall have put behind them the destructive uses of science and religion to accept the challenge that lies at the heart of both, which is for man to enter his heart and the nucleus of the atom and to harness from both the unlimited spiritual and physical resources to establish the golden age.[65]

scalar electromagnetics: Colloquial term for the electrodynamics that arise from considering transverse electromagnetic waves, longitudinal electromagnetic waves, time-polarized electromagnetic waves, electrogravitation, superluminal (faster than the speed of light) signals, interferometry, nonlinear optical functions, time-as-energy, and the infolded electrodynamics inside all usual electromagnetic fields, waves, and potentials. In secret superweapon projects, the Russian term for scalar electrodynamics is *energetics*.[66]

scalar wave: Characterized by magnitude only. With respect to polarization, however, a scalar photon is a term in use for a time-polarized photon, where the electromagnetic energy oscillates along the time axis.

The effects are observed as an oscillation in the rate of flow of time, hence a "time-density" oscillation. The term "scalar" with respect to polarization implies only that there is no vector component in three-dimensional space, even though a vector (and a variation of its magnitude) exists along the time axis. The effects are observed as an oscillation in the rate of flow of time, hence a "time-density" oscillation.[67]

Schrödinger, Erwin: A highly gifted German physicist with a broad education. After studying chemistry, he devoted himself for years to Italian painting. After this he took up botany, resulting in a series of papers on plant phylogeny. His great discovery, Schrödinger's wave equation, was made at the end of this epoch, during the first half of 1926. It came as a result of his dissatisfaction with the quantum condition in Niels Bohr's orbit theory and his belief that atomic spectra should really be determined by some kind of eigenvalue problem. For this work he shared with Paul Dirac the Nobel Prize in Physics in 1933.[68] *See also* Dirac, Paul.

second attention: First attention can be called the "knowing mind" and second attention can be called the "not knowing" or "unknowing mind." Second attention can be cultivated by relaxing the search for meaning. This can be experienced by relaxing the tendency to project and/or assume meaning onto whatever is perceived, in lieu of direct perception of the phenomena. This can also occur by dropping labels through an agreement to experience the world without naming what you are experiencing.[69]

serial processing: Processing that occurs sequentially. There is an explicit order in which operations occur, and the results of one action are generally known before a next action is considered.[70]

Sheldrake, Rupert: One of the world's most innovative developmental biologists, Sheldrake is best known for his theory of morphic fields and morphic resonance, which leads to a vision of a living, developing universe with its own inherent memory: "Over the course of fifteen years of research on plant development, I came to the conclusion that for understanding the development of plants, their morphogenesis, genes, and gene products are not enough. Morphogenesis also depends on organizing fields. The same arguments apply to the development of animals. Since the 1920s, many developmental biologists have proposed that biological organization depends on fields, variously called biological fields, or developmental fields, or positional fields, or morphogenetic fields."[71]

special relativity model: (1) The laws of physics are the same for all observers in uniform motion relative to one another (Galileo's principle of relativity), and (2) The speed of light in a vacuum is the same for all observers, regardless of their relative motion or of the motion of the source of the light.

spinor field: See torsion field.

super-coherent: More than a usual degree of coherence between the phases or two or more waves, so the interface effects may be produced between them, or a correlation between the phases or parts of a single wave.[72]

superposition: The simple linear addition and subtraction of two or more values, states, and so on. One of the key principles in field theories and in the concept of potentials. When the situation is sufficiently nonlinear, however, interaction of waves and potentials occurs instead of simple superposition.[73]

Tiller, William: Scientist and author, featured in the film *What the Bleep Do We Know?!*, fellow to the American Academy for the Advancement of Science, professor emeritus of Stanford University's Department of Materials Science, he spent thirty-four years in academia after nine years as an advisory physicist with the Westinghouse Research Laboratories. He has published more than 250 conventional scientific papers, three books, and several patents. In parallel, for more than thirty years, he has been avocationally pursuing serious experimental and theoretical study of the field of psychoenergetics, which will very likely become an integral part of "tomorrow's" physics. In this new area, he has published an additional one hundred scientific papers and four seminal books.[74]

time reverse: For electromagnetic waves, the process of forming a phase-conjugate wave. For a particle or a mass, the process of pumping the particle or mass with time-polarized electromagnetic waves so that the resident space-time curvature engine in the mass is amplified and phase-conjugated, forming a precise, amplified anti-engine, which then slowly time-reverses the mass back to a previous condition and state. We accent that *time-reversing a single object or single group of objects* is not the same thing as the "traveling into the past" so popularized by science fiction. For time travel to the past, the entire universe and everything in it except the traveler would have to be time-reversed. That would not seem possible by any stretch of the imagination today! On the other hand, time-reversal of a single thing, such as a particle or a wave—or even a group of things, such as a group of particles or waves—is not only feasible but readily achievable. A hole in a Dirac sea is a negative energy electron, for example, having negative energy and therefore negative mass prior to observation. After observation, it is seen as positive energy, a positive mass electron of opposite charge to the conventional electron and reacting with fields in opposite direction from the conventionally negatively charged electron. In short, after observation (interaction with mass) it becomes a positron.[75] *See also* Dirac sea; electron; phase-conjugate.

time-reversed wave: A phase-conjugate or time-reversed wave is a wave that travels backward through time. That is, it is capable of precisely retracing the path through space taken by another wave that traveled that path to a nonlinear mirror, stimulating the reflection of the time-reversed wave. Further, in retracing its invisible path through space, the phase-conjugate replica wave does not diverge as normal waves do. Instead, it continually converges upon its invisible trace.[76]

time travel: Time is illusory because past and future are connected to the present as possibilities. Our reality (that is, the universe) is a hologram, and at any instant, consciousness is the totality of the coherent signaling within the living matrix (including wave fronts reflected from specific information-containing structures that are maintained by our choices on a moment-to-moment basis and shape our experiences of the world). We each have valuable roles in the big picture—the matrix. Everything is connected to everything else holographically. In Matrix Energetics, we teach that photons can go backward and forward in time. They say that a wave of photons traveling forward in time represents the "advanced wave," and the one traveling backward in time is the "retarded wave." Where these phase-conjugated waves of photons intersect creates the present moment. Why do you suppose we have a part of the brain called the temporal lobe? Fred Alan Wolf, an author and independent scholar and researcher in physics and consciousness, theorizes that it might have something to do with time travel. He has said that there *is* a time-travel machine, and that it is our brain. In Matrix Energetics, a human body is ultimately composed of streams of photons held together by consciousness. If photons can travel backward and forward in time and that is ultimately what you are composed of, then this suggests that you can do so as well. The time-travel technique is based upon this fundamental concept.

torsion field: The quantum spin of empty space, the large-scale coherent effects of the spin of the particles in the virtual sea. Also known as a spinor field, axion field, spin field, and microlepton field. Initial work in this area of study was performed by Albert Einstein and Élie Cartan in the 1920s; now it is known as the ECT (Einstein-Cartan theory). Such fields are generated by classical spin or by the angular momentum density (on a macroscopic level) of any spinning object. The spinning of an object sets up polarization in two spatial cones, corresponding to a left torsion field and a right torsion field. At an atomic level, nuclear spin, as well as full atomic movements, may be the source of torsion fields, which means that all objects in nature generate their own torsion field.

These fields are not affected by distance, spread instantaneously in space, interact with material objects by exchanging information, and explain such phenomena as telepathy and photokinesis.[77]

Two-Point: A measuring tool that supercharges our ability to notice where we can plug into the grid of the *All That Is*. As you hold the two points, feel the connection between them. Feel and imagine that you are just working with photons or light. There is no body there, nothing solid except for your focus on the two points. You can imagine that you are linked and "entangled" with another person or area of yourself that you have chosen to focus on. Now, having connected those two points and realized that at the level of quantum physics the act of measuring actually causes a change in your measurement, you let go as if you were dropping a pebble into the pond. You imagine that you let go of the need for that to be physical and feel an expansive wave between those two points. When you practice the art of Two-Point, it represents a new paradigm for things you can do or access with your sensory modality of touch. If you endeavor to do this on a daily basis, you'll begin to have glimpses of the hidden reality and its complexities behind the shroud of daily events. Things no longer happen to you. Instead, you begin to take responsibility for your creative use of universal energy.

uncertainty principle: The principle of quantum mechanics, formulated by Werner Heisenberg, that the accurate measurement of one of two related, observable quantities, as position and momentum or energy and time, produces uncertainties in the measurement of the other, such that the product of the uncertainties of both quantities is equal to or greater than $h/2\pi$, where h equals Planck's constant. Also called the indeterminacy principle (*Random House Dictionary* [New York: Random House, 2009]). In conventional quantum physics, the origin of zero-point energy is the Heisenberg uncertainty principle since a parallel uncertainty exists between measurements involving time and energy (and other so-called conjugate variables in quantum mechanics). This minimum uncertainty is not due to any correctable flaws in measurement, but rather reflects an intrinsic quantum fuzziness in the very nature of energy and matter springing from the wave nature of the various quantum fields. This leads to the concept of zero-point energy.[78]

unified theory: A unified theory of the four forces of physics—the electromagnetic, gravitational, strong, and weak forces—that is not just an intellectual model but also engineerable on the laboratory bench and in actual physical systems using higher-symmetry electrodynamics as a special subset of Mendel Sachs's unified field theory.[79]

vacuum: Space devoid of observable matter. In modern theory, the "empty" space is in fact bristling with very rapid fluctuations of electromagnetic energy, remaining in the virtual state. It is also filled with a violent, fluctuating flux of virtual particles, appearing and disappearing so quickly that an individual particle does not persist long enough to be individually detected. Thus, the vacuum is extraordinarily energetic, but the energy is in very special form (fleeting violent fluctuations and virtual particle flux). Nonetheless, because it contains enormous energy, the average vacuum may be considered a potential.[80]

vector: In mathematics, a quantity having both magnitude and direction. For example, an ordinary quantity, or scalar, can be exemplified by the distance 6 km; a vector quantity can be exemplified by the term 6 km north. Vectors are usually represented by directed line segments; the length of the line segment is a measure of the vector quantity, and its direction is the same as that of the vector.[81]

virtual particle: A fleeting quantum particle that spontaneously appears and disappears so swiftly that it cannot be individually observed; it exists only temporarily. The virtual particle does not satisfy the usual relation between energy, momentum, and mass because it is underneath the Heisenberg uncertainty principle. The virtual particle can have any amount of energy momentarily, so long as the product of its energy and the time interval of its existence is less than the uncertainty principle's minimum magnitude. Nonetheless, the interactions of large numbers of virtual particles with a mass or charge can combine to generate real observable effects. In quantum field theory, all forces of nature are caused by the interaction of the forced mass entity with virtual particles.[82]

virtual reality: Virtual realities are consciousness technologies that allow users to interact with an energetic set of patterns or environment, be it a real or imagined one. If everything is a result of the interplay of consciousness with the foundation blocks of physical matter, photons, and virtual particles, then in one sense every view of reality is virtual and not actual.

virtual reality template: A pattern of data that may be used to match input or generate output into a virtual reality system.

visible electromagnetic spectrum: Visible light waves are the only electromagnetic waves we can see. We see these waves as the colors of the rainbow. Each color has a different wavelength. Red has the longest wavelength and violet has the shortest wavelength. When all the waves are seen together, they make white light. When white light shines through a prism, the white light is broken apart into the colors of the visible light spectrum. Water vapor in the atmosphere can also break apart wavelengths, creating a rainbow.[83]

waveform: A graphical representation of the shape of a wave for a given instant in time over a specified region in space.[84]

weak quantum measurement: Weak measurements are a type of quantum measurement, where the measured system is weakly coupled to the measuring device so that the system is not disturbed by the measurement. Although seemingly contradictory to some basic aspects of quantum theory, the formalism lies within the boundaries of the theory and does not contradict any fundamental concept. The idea of weak measurements and weak values, first developed by Yakir

Aharonov, David Albert, and Lev Vaidman, is especially useful for gaining information about pre- and post-selected systems described by the two-state vector formalism.[85]

Wheeler, John Archibald: An eminent American theoretical physicist who coined the terms "black hole" and "wormhole" and the phrase "it from bit," and whose myriad scientific contributions figured in many of the research advances of the twentieth century. Wheeler was known for his drive to address big, overarching questions in physics, subjects he liked to say merged with philosophical questions about the origin of matter, information, and the universe. He was a young contemporary of Albert Einstein and Niels Bohr and was a driving force in the development of both the atomic and hydrogen bombs, and in later years he became the father of modern general relativity.[86]

zero-point energy: The energy that remains when all other energy is removed from a system. This behavior is demonstrated, for example, by liquid helium. As the temperature is lowered to absolute zero, helium remains a liquid, rather than freezing to a solid, owing to the irremovable zero-point energy of its atomic motions. (Increasing the pressure to twenty-five atmospheres causes helium to freeze.) Quantum mechanics predicts the existence of what are usually called "zero-point" energies for strong, weak, and electromagnetic interactions, where the term refers to the energy of the system at temperature $T = 0$, or the lowest quantized energy level of a quantum mechanical system. Turning to Werner Heisenberg's uncertainty principle, one finds that the lifetime of a given zero-point photon, viewed as a wave, corresponds to an average distance traveled of only a fraction of its wavelength. Such a wave "fragment" is somewhat different than an ordinary plane wave, and it is difficult to know how to interpret this.[87]

zero-point energy field: In quantum field theory, electromagnetic radiation can be pictured as waves flowing through space at the speed of light. The waves are not waves of anything substantive but ripples in a state of a theoretically defined field. These waves do carry energy and momentum, however, and each has a specific direction, frequency, and polarization state. Each wave represents a "propagating mode of the electromagnetic field." The zero-point field is the lowest energy state of a field, its lowest ground state not equal to zero. Quantum physics predicts that all space must be filled with electromagnetic zero-point fluctuations, thus creating a universal sea of zero-point energy. This phenomenon gives the quantum vacuum a complex structure, which can be probed experimentally, via the Casimir effect, for example. The term *zero-point field* is sometimes used as a synonym for the vacuum state of an individual quantized field. The electromagnetic zero-point field is loosely considered as a sea of background electromagnetic energy that fills the vacuum of space.[88] *See also* Casimir effect.

zero-point reference time reference: It is one thing to make invisible (or time travel) an object like a ship, but people have individual time references. You can "reference" an object to "travel," but with people you have to individually reference their "reference points" to make sure it will all work.[89]

NOTES

Introduction

1. Rupert Sheldrake, www.sheldrake.org/Resources/glossary.*

2. Richard Bartlett, *Matrix Energetics: The Science and Art of Transformation* (Hillsboro, OR: Atria Books/Beyond Words Publishing, 2007).

Chapter 4

1. The Internet Movie Database, http://www.imdb.com/character/ch0001072/quotes.*

2. The Quotation Page, published by Michael Moncur, http://www.quotationspage.com/quotes/Wernher_von_Braun/.*

Chapter 6

Original definition by Dr. Mark Dunn and Dr. Richard Bartlett.

Chapter 7

1. "The shell game that we play . . . is technically called 'renormalization.' But no matter how clever the word, it is still what I would call a dippy process! Having to resort to such hocus-pocus has prevented us from proving that the theory of quantum electrodynamics is mathematically

All URL sources accessed spring 2009.

253

self-consistent. It's surprising that the theory still hasn't been proved self-consistent one way or the other by now; I suspect that renormalization is not mathematically legitimate." Richard P. Feynman, *QED: The Strange Theory of Light and Matter* (New York: Penguin, 1990), 128.

2. Tony Rothman, *Everything's Relative: And Other Fables from Science and Technology* (Hoboken, NJ: John Wiley & Sons, 2003), 78–84.

3. P. Halmos, "The Legend of John von Neumann," *American Mathematical Monthly* (April 1973), 382–94.

4. "Einstein's postulates: (1) All the laws of physics are equally valid in all inertial frames of reference; (2) the speed of light is the same to every inertial observer; and (3) the observable local effects of a gravitational field are indistinguishable from those arising from the acceleration of the frame of reference. The first is called the special relativity principle, the second is called the law of light propagation, and the third principle is called the equivalence principle." Thomas E. Bearden, *Energy from the Vacuum: Concepts and Principles* (Santa Barbara, CA: Cheniere, 2002), 647.

5. David Smith, *Quantum Sorcery* (Tokyo, Japan: Konton Publishing, 2006), 37.

6. The Internet Movie Database, http://www.imdb.com/title/tt0087 332/quotes.*

7. Indigenous Weather Modification (TWM) is a site dedicated to teaching indigenous WM technology, http://twm.co.nz/forbquco.html.*

8. See the interview on Quantum Tantra with Nick Herbert, conducted by Joseph Matheny, http://74.125.155.132/search?q=cache: s59QayXJ2oAJ:www.incunabula.org/inc3.html+ong%27s+hat+joseph+ matheny+atoms+are+things&cd=1&hl=en&ct=clnk&gl=us.*

9. Edward Whitmont, *The Alchemy of Healing: Psyche and Soma* (Berkeley, CA: North Atlantic Books, 1996), 29, quoted in Michael Talbot, *Beyond the Quantum* (New York: Bantam Books, 1988), 155.

10. Dennis Overbye, "John A. Wheeler; Physicist Who Coined the Term 'Black Hole,' Is Dead at 96," *New York Times* (April 14, 2008), http://www .nytimes.com/2008/04/14/science/14wheeler.html?_r=1&pagewanted=2.*

All URL sources accessed spring 2009.

Chapter 9

1. The Internet Movie Database, http://www.imdb.com/character/ ch0007463/quotes.

2. Quotes, sayings, and poem at Litera.co.uk., shttp://www.litera.co .uk/author/jim_morrison/.*

Chapter 10

1. Paramahansa Yogananda, *Autobiography of a Yogi* (Los Angeles: Self-Realization Fellowship, 1946), 320–32.

Chapter 11

1. Richard Bartlett, "The Music of Your Mind," radio interview by Marla Frees, Portland, Oregon, May 3, 2007.

Chapter 14

1. Hector Garcia, guest lecturer, Matrix Energetics seminar, May 2008.

Chapter 16

1. David Hatcher, *Antigravity and the World Grid* (Kempton, IL: Adventures Unlimited, 2006), 112.

2. Morris K. Jessup and Carlos Allende, *The Allende Letters and the VARO Edition of The Case For the UFOs* (New Brunswick, New Jersey: Global Communications/Conspiracy Journal, 2007), 28. Also see Charles Berlitz, interviewed by Dr. J. Manson Valentine, http://www.scribd.com/ doc/13355366/The-Philadelphia-Experiment-Charles-Berlitz.*

3. Alexandra Bruce, *The Philadelphia Experiment Murder: Parallel Universes and the Physics of Insanity* (New York: Sky Books, 2001), 158.

4. Ibid., 157.

5. Ibid., 159.

6. Ibid., 160–161.

7. Michio Kaku, *Physics of the Impossible* (New York: Doubleday Random House, 2008), 48.

*All URL sources accessed spring 2009.

8. Tachi-Kawakami Laboratory Graduate School of Information Science and Technology, The University of Tokyo, http://tachilab.org.*

9. Sarah Yang, Media Relations, "Invisibility Shields One Step Closer with New Metamaterials that Bend Light Backwards," *UC Berkeley News*, August 11, 2008, http://berkeley.edu/news/media/releases/2008/08/11_light.shtml.*

10. Kaku, *Physics of the Impossible*, 38.

11. Yang, "Invisibility Shields One Step Closer with New Metamaterials that Bend Light Backwards."*

12 Kaku, *Physics of the Impossible*, 38.

13. Jay Alfred, *Between the Moon and Earth* (Victoria, BC: Trafford Publishing, 2006), 31.

14. Plasma Universe presented by Los Alamos National Laboratory, associated with the IEEE Nuclear and Plasma Sciences Society, http://plasmascience.net/tpu/ubiquitous.html.*

15. Magnet Import, http://www.magnetimport.no/subtle.html.*

16. Jay Alfred, "Bioplasma Bodies: The Ovoid or the Body's Magnetosphere," Ezineartics, 2007. http://ezinearticles.com/?Bioplasma-Bodies---The-Ovoid-or-the-Bodys-Magnetosphere&id=770297 (emphasis mine).*

17. Steve Richards, *Invisibility: Mastering the Art of Vanishing* (London: Aquarian Press, 1982), 16–17.

18. Ibid., 41.

19. Mark L. Prophet and Elizabeth Clare Prophet, *Saint Germain on Alchemy: Formulas for Self-Transformation* (Livingston, MT: Summit University Press, 1988), 200.

Chapter 18

1. Comte de Saint Germain and Mark Prophet, *Studies in Alchemy: The Science of Self-Transformation* (Gardiner, MT: Summit University Press, 1997), http://www.summituniversitypress.com/books/sgalchemy.html.*

2. "Where there is no vision, the people perish: but he that keeps the law, happy is he." Proverbs 29:18 (American King James Version).

3. Matthew 7:7 (American King James Version).

Chapter 20

1. Mark L. Prophet, *Science of the Spoken Word*, 8th ed. (Gardiner, MT: Summit University Press, 1998). First attention expresses a function of physical sight and intellect; second attention conveys a function of the energetic body.

2. "Thou shalt also decree a thing, and it shall be established unto thee; and light shall shine upon thy ways." Job 22:28 (American Standard Version).

Chapter 23

1. Mervin Rees discovered the temporal sphenoidal (TS) line, reflex points points for muscles, organs, and glands, during World War II. In 1955 he began practicing in Sedan, Kansas, and in 1956, he began the sacro occipital technique. In 1974 Rees was installed as a director of Sacro Occipital Research Society International, and in 1980 he introduced his original technique, the Soft Tissue Orthopedic technique, which eventually led to a harmonic technique. See Maeda Shigeru, "Chiropractic in Japan," 1996, http://www.asahi-net.or.jp/~xf6s-med/eover view.html. Also see the Association Culturelle Chiropractic Team, "The First European AK Meetings, 1976–1978," *The International Journal of Applied Kinesiology and Kinesiologic Medicine*, Issue 20 (Fall 2005), http://www.kinmed.com*

Chapter 24

1. Matthew 13:12 (American Standard Version).

2. Sheldrake, www.sheldrake.org/Resources/glossary.*

3. Thomas E. Bearden, *Radionics: Action at a Distance*, DVD, (1990; Atlanta, GA: Cheniere Media, 2006), http://www.cheniere.org/sales/buy-ra.htm.*

Matrix Glossary

1. The Light Party, "Radionics" (1996), www.lightparty.com/Health/Radionics.html.*

2. *The American Heritage Dictionary of the English Language*, 4th ed. s.v. "aether physics"; *Encyclopedia Britannica Online*, s.v. "aether phyics".*

3. Thomas E. Bearden (1997), http://www.cheniere.org/techpapers/Annotated%20Glossary.htm.*

4. *Merriam-Webster's Collegiate Dictionary*, 11th ed., s.v. "archetypes"; Mike Adams, ed., "Survey Results Reveal the Most Trusted Health News Websites and Personalities," NaturalNews.com (April 9, 2008), http://www.naturalnews.com/.*

5. E. T. Whittaker, "On the Partial Differential Equations of Mathematical Physics," *Mathematische Annalen*, 57 (1903), 333–355; Bearden, http://www.cheniere.org/techpapers/Annotated%20Glossary.htm.*

6. Bearden, http://www.cheniere.org/books/excalibur/glossary/014edited.htm.*

7. Institute for Bioelectromagnetics and New Biology, http://www.bion.si.*

8. Marshall Space Flight Center, Biography of Dr. Wernher von Braun, http://history.msfc.nasa.gov/vonbraun/bio.html.*

9. Nobelprize.org, "Biography," http://nobelprize.org/nobel_prizes/physics/laureates/1929/broglie-bio.html.*

10. Thomas E. Bearden, *Energy from the Vacuum: Concepts and Principles* (Santa Barbara, CA: Cheniere, 2002), 622.

11. Calphysics Institute, "Zero-Point Energy and Zero Point Field," http://www.calphysics.org/zpe.html.*

12. Bearden, *Energy from the Vacuum*, 626.

13. *Dictionary of Science and Technology*, 1st ed., Christopher G. Morris, ed, s.v. "coherence."

14. Claus Kiefer, "On the Interpretation of Quantum Theory: From Copenhagen to the Present Day" (Oct. 2002), http://arxiv.org/abs/quant-ph/0210152.*

15. David E. Joyce, "Dave's Short Course on Complex Numbers: Reciprocals, Conjugates, and Division" (1999), http://www.clarku.edu/~djoyce/complex/div.html.*

16. Erich Joos, *Decoherence* (2008), http://www.decoherence.de/.*

17. Nobelprize.org, "Biography," http://nobelprize.org/nobel_prizes/physics/laureates/1933/dirac-bio.html.*

18. Calphysics Institute, "Zero-Point Energy and Zero Point Field," http://www.calphysics.org/zpe.html; posted comment on "Dirac's Hidden Geometry," blog thread *Not Even Wrong* (Sept. 25, 2005), http://www.math.columbia.edu/~woit/wordpress/?p=262#comment-5066.*

19. Bearden, *Energy from the Vacuum*, 660.

20. Ibid., 65.

21. Richard Feynman, *The American Heritage Dictionary of the English Language*, 4th ed. s.v. "Feynman, Richard."

22. William A. Tiller, *Science and Human Transformation: Subtle Energies, Intentionality, and Consciousness* (Walnut Creek, CA: Pavior Publishing, 1997), 89.

23. Garcia Chiropractic Holistic Center (2009), http://www.garcia holisticchiro.com/about_dr.php.*

24. Bearden, *Energy from the Vacuum*, 677.

25. Ibid., 678.

26. William Reville, prof., "Ireland's Greatest Mathematician," University College, Cork, Ireland (2004), http://understandingscience.ucc.ie/pages/sci_williamrowanhamilton.htm.*

27. Institute of HeartMath, "Science of the Heart: Exploring the Role of the Heart in Human Performance" (2009), 4, http://www.heart-math.org/research/science-of-the-heart-head-heart-interactions.html.*

28. Bearden, *Energy from the Vacuum*, 680.

29. Nobelprize.org, "Biography," http://nobelprize.org/nobel_prizes/physics/laureates/1932/heisenberg-bio.html.*

30. Boston University School of Theology: Anna Howard Shaw Center, "Biography," http://sthweb.bu.edu/shaw/anna-howard-shaw center/biography?view=mediawiki&article=Nick_Herbert_%28 physicist%29.*

31. Hologram: Michael Talbot, *The Holographic Universe* (New York: HarperCollins, 1991), 14.

32. Bearden, *Energy from the Vacuum*, 719.

33. Glenda Green, "More Than Meets the Eye," www.glendagreen .com.*

34. Fred Alan Wolf, *The Eagle's Quest: A Physicist Finds the Scientific Truth at the Heart of the Shamanic World* (New York: Touchstone, 1997), 145–46.

35. Danish mathematician Tor Nørretranders, quoted in James Oschman, *Energy Medicine in Therapeutics and Human Performance* (Boston: Butterworth-Heinemann, 2003); "Ponderings and Learnings," http://www.craniosacralpath.com/blog.*

36. Stanford Encyclopedia of Philosophy, "Many-Worlds Interpretation of Quantum Mechanics" (Mar. 24, 2002), http://plato.stanford .edu/entries/qm-manyworlds/.*

37. James Clerk Maxwell Foundation, http://www.clerkmaxwell foundation.org/html/who_was_maxwell_.html.*

38. Bearden, *Energy from the Vacuum*, 693.

39. Bearden, http://www.cheniere.org/techpapers/Annotated%20 Glossary.htm.*

40. Joe McMoneagle, "Business Bio," http://blog.mceagle.com/ about/joe-bio-biz.*

41. Sheldrake, www.sheldrake.org/Resources/glossary.*

42. Ibid.

43. Ibid.

44. Bearden, *Energy from the Vacuum*, 694.

45. Thomas E. Bearden, *AIDS Biological Warfare* (Greenville, TX: Tesla Book Company, 1988), 152.

46. Bearden, *Energy from the Vacuum*, 663.

47. MSN Encyclopedia Article Center, http://encarta.msn.com/encyclopedia_761579159/John_Von_Neumann.html.*

48. Carlos Castaneda, *Journey to Ixtlan* (New York: Washington Square Press, 1972), 189.

49. Bearden, http://www.cheniere.org/techpapers/Annotated%20Glossary.htm.*

50. Bearden, *Energy from the Vacuum*, 699.

51. Ibid., 719.

52. Josh Clark, "Do Parallel Universes Really Exist?", *How Stuff Works,* http://science.howstuffworks.com/parallel-universe.htm.*

53. Bearden, http://www.cheniere.org/techpapers/Annotated %20 Glossary.htm.*

54. Frank Swain, SciencePunk.com (Oct. 5, 2006), http://www.sciencepunk.com/2006/10/albert-abrams-2/.*

55. Bearden, *Energy from the Vacuum*, 703.

56. Thomas J. McFarlane, *Quantum Physics, Depth Psychology, and Beyond,* The Center of Integral Science (June 21, 2000), http://www.integralscience.org/psyche-physis.html.*

57. Richard P. Feynman, Robert B. Leighton, Matthew Sands, *The Feynman Lectures on Physics*, vol. 1, 2nd ed. (London: Addison Wesley, 2005), 11.

58. United States Psychotronics Association, "What Is Psychotronics?," http://www.psychotronics.org/aboutus.htm.*

59. *Dictionary of Science and Technology*, s.v. "quantum field theory."

60. Bearden, *Energy from the Vacuum*, 710.

61. Ibid., 711.

62. Ibid.

63. *Encyclopaedia Britannica Online*, s.v. "renormalization."

All URL sources accessed spring 2009.

64. Bruce Lee Foundation, http://www.bruceleefoundation.com/index1000.html.*

65. The Summit Lighthouse, Summit University, "Ascended Masters," http://www.tsl.org/Masters/SaintGermain.asp.*

66. Bearden, *Energy from the Vacuum*, 714.

67. Ibid.

68. Nobelprize.org, "Biography," http://nobelprize.org/nobel_prizes/physics/laureates/1933/schrodinger-bio.html.*

69. Antero Alli, ParaTheatrical Research, http://www.paratheatrical.com.*

70. Artificial Intelligence Lab, University of Michigan, http://ai.eecs.umich.edu/cogarch0/common/prop/serial.html.*

71. Sheldrake, www.sheldrake.org/papers/Morphic/morphic_intro.html; www.sheldrake.org/homepage.html; www.sheldrake.org/About/biography/pwfund.html.*

72. *Dictionary of Science and Technology*, s.v. "super-coherent."

73. Bearden, *Energy from the Vacuum*, 722.

74. William A. Tiller Foundation, "bio," http://www.tillerfoundation.com/biography.php.*

75. Bearden, *Energy from the Vacuum*, 726.

76. Bearden, *AIDS Biological Warfare*, 105.

77. Patent Storm, U.S. Patent 6548752: system and method for generating a torsion field, issued April 15, 2003, http://www.patentstorm.us/patents/6548752/description.html; Uvitor, "history," http://www.shipov.com/history.html.*

78. *Random House Dictionary*, 4th ed., s.v. "uncertainty principle"; Calphysics Institute, "Zero-Point Energy and Zero Point Field," http://www.calphysics.org/zpe.html; Stanford Encyclopedia of Philosophy, "The Uncertainty Princple," (Jul. 3, 2006), http://plato.stanford.edu/entries/qt-uncertainty/.*

79. Bearden, *Energy from the Vacuum*, 728.

80. Ibid., 729.

81. MSN Encyclopedia Article Center, http://encarta.msn.com/encyclopedia_761572843/Vector_(mathematics).html.*

82. Bearden, http://www.cheniere.org/techpapers/ Annotated%20 Glossary.htm.*

83. National Aeronautics and Space Administration, March 27, 2007, "Visible Light Waves," http://science.hq.nasa.gov/kids/imagers/ms/visible.html.*

84. *Dictionary of Science and Technology*, s.v. "waveforms."

85. Yakir Aharonov, D. Z. Albert, and Lev Vaidman, "How the Result of a Measurement of a Component of the Spin of a Spin-1/2 Particle Can Turn Out to Be 100," *Physical Review Letters*, 1988.

86. Kitta MacPherson, "Leading Physicist John Wheeler Dies at Age 96," *News at Princeton University* (April 14, 2008), http://www.princeton.edu/main/news/archive/S20/82/08G77/.*

87. Calphysics Institute, "Zero-Point Energy and Zero Point Field," http://www.calphysics.org/zpe.html.*

88. Ibid.

89. The Philadelphia Experiment, http://www.phils.com.au/philadelphia.htm.*

SELECTED BIBLIOGRAPHY

Alfred, Jay. *Between the Moon and Earth*. Victoria, BC: Trafford Publishing, 2006.

———. *Brains and Realities*. Victoria, BC: Trafford Publishing, 2006.

———. *Our Invisible Bodies: Scientific Evidence for Subtle Bodies*. Victoria, BC: Trafford Publishing, 2006.

Aspden, Harold. *Modern Aether Science*. Southampton, UK: Sabberton Publications, 1972.

Bartlett, Richard. *Matrix Energetics: The Science and Art of Transformation*. Hillsboro, OR: Atria Books/Beyond Words, 2007.

Bearden, Thomas E. *AIDS Biological Warfare*. Greenville, TX: Tesla Book Company, 1988.

———. *Excalibur Briefing: Explaining Paranormal Phenomena*. Santa Barbara, CA: Cheniere, 2002.

———. *Energy from the Vacuum: Concepts and Principles*. Santa Barbara, CA: Cheniere, 2002.

———. *Oblivion: America at the Brink*. Santa Barbara, CA: Cheniere, 2005.

———. *Fer de Lance*. Santa Barbara, CA: Cheniere, 2003.

———. *Gravitobiology: A New Biophysics*. Santa Barbara, CA: Cheniere, 2003.

Bedini, John, and Thomas Bearden. *Free Energy Generation—Circuits and Schematics: 20 Bedini-Bearden Years*. Santa Barbara, CA: Cheniere, 2006.

Bentov, Itzhak. *Stalking the Wild Pendulum: On the Mechanics of Consciousness*. Rochester, VT: Destiny Books, 1988.

———. *A Brief Tour of Higher Consciousness: A Cosmic Book on the Mechanics of Creation*. Rochester, VT: Destiny Books, 2006.

Cathie, Bruce L. *The Harmonic Conquest of Space*. Kempton, IL: Adventures Unlimited, 1998.

———. *The Energy Grid*. Kempton, IL: Adventures Unlimited, 1997.

Cheney, Margaret. *Tesla: Man Out of Time*. New York: Barnes & Noble Books, 1993.

Childress, David Hatcher. *Anti-Gravity and the Unified Field*. Kempton, IL: Adventures Unlimited, 2001.

———. *The Time Travel Handbook: A Manual of Practical Teleportation and Time Travel*. Kempton, IL: Adventures Unlimited, 1999.

Chopra, Deepak. *The Third Jesus: The Christ We Cannot Ignore*. New York: Harmony, 2008.

Coats, Callum. *Living Energies: An Exposition of Concepts Related to the Theories of Viktor Schauberger*. Dublin, Ireland: Gateway Books, 2001.

Cook, Nick. *The Hunt for Zero Point: One Man's Journey to Discover the Biggest Secret Since the Invention of the Atom Bomb*. London: Century, 2001.

Dalal, A. S. *Powers Within*. Pondicherry, India: Sri Aurobindo Ashram Publications Department, 1999.

Deary, Terry. *Vanished!* Boston: Kingfisher, 2004.

Dennett, Preston. *Human Levitation: A True History and How-to Manual*. Grand Rapids, MI: Schiffer Publishing, 2006.

Dolley, Chris. *Shift*. Riverdale, NY: Baen Books, 2007.

Dowling, Levi. *The Aquarian Gospel of Jesus the Christ*. New York: Cosimo Classics, 2007.

Dunn, Christopher. *The Giza Power Plant: Technologies of Ancient Egypt.* Rochester, VT: Bear & Company, 1998.

Durr, Hans-Peter, Fritz-Albert Popp, and Wolfram Schommers. *What Is Life? Scientific Approaches and Philosophical Positions.* Hackensack, NJ: World Scientific, 2002.

Edwards, Harry. *Harry Edwards: Thirty Years a Spiritual Healer.* Surrey, UK: Jenkins, 1968.

Farrell, Joseph P. *The Cosmic War: Interplanetary Warfare, Modern Physics, and Ancient Texts.* Kempton, IL: Adventures Unlimited, 2007.

———. *The Giza Death Star Deployed: The Physics and Engineering of the Great Pyramid.* Kempton, IL: Adventures Unlimited, 2003.

———. *The Giza Death Star Destroyed: The Ancient War for Future Science.* Kempton, IL: Adventures Unlimited, 2005.

———. *Reich of the Black Sun: Nazi Secret Weapons & the Cold War Allied Legend.* Kempton, IL: Adventures Unlimited, 2005.

———. *Secrets of the Unified Field: The Philadelphia Experiment, the Nazi Bell, and the Discarded Theory.* Kempton, IL: Adventures Unlimited, 2008.

———. *The SS Brotherhood of the Bell: The Nazis' Incredible Secret Technology.* Kempton, IL: Adventures Unlimited, 2006.

Friedman, Norman. *The Hidden Domain: Home of the Quantum Wave Function, Nature's Creative Source.* Eugene, OR: Woodbridge Group, 1997.

Garrison, Cal. *Slim Spurling's Universe: The Light-Life Technology: Ancient Science Rediscovered to Restore the Health of the Environment and Mankind.* Frederick, CO: IX-EL Publishing, 2004.

Green, Glenda. *The Keys of Jeshua.* Sedona, AZ: Spiritis Publishing, 2004.

Harbison, W. A. *Projekt UFO: The Case for Man-made Flying Saucers.* Charleston, SC: BookSurge, 2007.

Harpur, Patrick. *Daimonic Reality: A Field Guide to the Otherworld.* Ravensdale, WA: Pine Winds, 2003.

Ho, Mae-Wan. *The Rainbow and the Worm: The Physics of Organisms.* Hackensack, NJ: World Scientific, 1998.

Hoagland, Richard C., and Mike Bara. *Dark Mission: The Secret History of NASA*. Los Angeles: Feral House, 2007.

James, John. *The Great Field: Soul at Play in a Conscious Universe*. Fulton, CA: Energy Psychology Press, 2008.

King, Moray B. *The Energy Machine of T. Henry Moray: Zero-Point Energy & Pulsed Plasma Physics*. Kempton, IL: Adventures Unlimited, 2005.

Knight, Christopher, and Alan Butler. *Who Built the Moon?* London: Watkins Publishing, 2005.

Kraft, Dean. *A Touch of Hope: A Hands-On Healer Shares the Miraculous Power of Touch*. New York: Berkley Trade, 1998.

Kron, Gabriel. *Tensors for Circuits*. New York: Dover Publications, 1959.

Laszlo, Ervin. *Science and the Akashic Field: An Integral Theory of Everything*. Rochester, VT: Inner Traditions, 2004.

———. *Science and the Reenchantment of the Cosmos: The Rise of the Integral Vision of Reality*. Rochester, VT: Inner Traditions, 2006.

LaViolette, Paul A. *Genesis of the Cosmos: The Ancient Science of Continuous Creation*. Rochester, VT: Bear & Company, 2004.

———. *Secrets of Antigravity Propulsion: Tesla, UFOs, and Classified Aerospace Technology*. Rochester, VT: Bear & Company, 2008.

———. *Subquantum Kinetics: A Systems Approach to Physics and Cosmology*. Alexandria, VA: Starlane Publications, 2003.

Lilly, John C. *The Scientist: A Metaphysical Autobiography*. Oakland, CA: Ronin Publishing, 1996.

Lloyd, Seth. *Programming the Universe: A Quantum Computer Scientist Takes on the Cosmos*. London: Vintage Books, 2007.

Lyne, William R. *Pentagon Aliens*. Lamy, NM: Creatopia Productions, 1999.

Maxwell, James Clerk. *An Elementary Treatise on Electricity*. Mineola, NY: Dover Publications, 2005.

Monroe, Robert A. *Journeys Out of the Body*. Garden City, NY: Anchor, 1977.

Moore, William, and Charles Berlitz. *The Philadelphia Experiment: Project Invisibility*. New York: Fawcett, 1995.

Murakami, Aaron C. *The Quantum Key*. Seattle: White Dragon, 2007.

Oschman, James L. *Energy Medicine: The Scientific Basis*. New York: Churchill Livingstone, 2000.

Pickover, Clifford A. *Sex, Drugs, Einstein, and Elves: Sushi, Psychedelics, Parallel Universes, and the Quest for Transcendence*. Petaluma, CA: Smart Publications, 2005.

Popp, Fritz Albert, and L. V. Belousov. *Integrative Biophysics: Biophotonics*. New York: Springer, 2003.

Prophet, Mark L., and Elizabeth Clare Prophet. *Saint Germain on Alchemy: Formulas for Self-Transformation*. Livingston, MT: Summit University, 1993.

Randles, Jenny. *Time Travel: Fact, Fiction & Possibility*. New York: Blandford Press, 1994.

Regardie, Israel. *The Golden Dawn: The Original Account of the Teachings, Rites & Ceremonies of the Hermetic Order*. St. Paul, MN: Llewellyn Publications, 1986.

Richards, Steve. *Invisibility: Mastering the Art of Vanishing*. Wellingborough, UK: Aquarian Press, 1982.

Rothman, Tony. *Everything's Relative: And Other Fables from Science and Technology*. Hoboken, NJ: John Wiley & Sons, 2003.

Rothman, Tony, and George Sudarshan. *Doubt and Certainty*. Reading, MA: Helix Books, 1998.

Russell, Edward W. *Report on Radionics: The Science Which Can Cure Where Orthodox Medicine Fails*. Essex, UK: C. W. Daniel, 1973.

Russell, Ronald, and Charles T. Tart. *The Journey of Robert Monroe: From Out-of-Body Explorer to Consciousness Pioneer*. Charlottesville, VA: Hampton Roads Publishing, 2007.

Samanta-Laughton, Manjir. *Punk Science: Inside the Mind of God*. Ropley, Hants, UK: O Books, 2006.

Sheldrake, Rupert. *The Presence of the Past: Morphic Resonance and the Habits of Nature*. Rochester, VT: Park Street, 1988.

Scheinfeld, Robert. *Busting Loose from the Money Game*. Hoboken, NJ: Wiley, 2006.

Strauss, Michael. *Requiem for Relativity: The Collapse of Special Relativity*. Pembroke Pines, FL: RelativityCollapse.com, 2004.

Sussman, Janet I. *Timeshift: The Experience of Dimensional Change*. Fairfield, IA: Time Portal Publications, 1996.

Swanson, Claude. *The Synchronized Universe: New Science of the Paranormal*. Tucson, AZ: Poseidia Press, 2003.

Talbot, Michael. *Mysticism and the New Physics*. New York: Penguin, 1993.

Tansley, David V. *Radionics Interface with the Ether Fields*. Boston: C. W. Daniel, 1975.

Tiller, William A. *Science and Human Transformation: Subtle Energies, Intentionality, and Consciousness*. Walnut Creek, CA: Pavior Publishing, 1997.

Tiller, William A., Walter Dibble, and Gregory J. Fandel. *Some Science Adventures with Real Magic*. Walnut Creek, CA: Pavior Publishing, 2005.

Tiller, William A., Walter Dibble, and Michael Kohane. *Conscious Acts of Creation: The Emergence of a New Physics*. Walnut Creek, CA: Pavior Publishing, 2001.

Valone, Thomas F. *Electrogravitics II: Validating Reports on a New Propulsion Methodology*. Washington, DC: Integrity Research Institute, 2000.

———. *Harnessing the Wheelwork of Nature: Tesla's Science of Energy*. Kempton, IL: Adventures Unlimited, 2002.

———. *Practical Conversion of Zero-Point Energy: Feasibility Study of the Extraction of Zero-Point Energy from the Quantum Vacuum for the Performance of Useful Work*. 3rd ed. Beltsville, MD: Integrity Research Institute, 2003.

———. *Zero Point Energy: The Fuel of the Future*. Beltsville, MD: Integrity Research Institute, 2007.

Valone, Thomas F., and Elizabeth A. Rausher. *Electrogravitics Systems: Reports on a New Propulsion Methodology*. Washington, DC: Integrity Research Institute, 2001.

Violette, John R. *Extra-Dimensional Universe: Where the Paranormal Becomes Normal.* Charlottesville, VA: Hampton Roads, 2005.

Wang, Robert. *The Qabalistic Tarot: A Textbook of Mystical Philosophy.* Columbia, MD: Marcus Aurelius Press, 2004.

Wesson, Paul S. *Five-Dimensional Physics: Classical and Quantum Consequences of Kaluza-Klein Cosmology.* Hackensack, NJ: World Scientific, 2006.

Yogananda, Paramahansa. *The Second Coming of Christ: The Resurrection of the Christ Within You.* Los Angeles: Self-Realization Fellowship, 2004.
———. *Self-Realization.* Los Angeles: Self-Realization Fellowship, 2004.
———. *The Yoga of Jesus: Understanding the Hidden Teachings of the Gospels.* Los Angeles: Self-Realization Fellowship, 2007.

the jewish princess cookbook

having your cake & eating it...

Georgie Tarn & Tracey Fine

quadrille

This paperback edition first published in 2007 by
Quadrille Publishing Limited
Alhambra House
27-31 Charing Cross Road
London WC2H 0LS

editorial director Jane O'Shea
project editor Jamie Ambrose
designer Claire Peters
production Funsho Asemota

Cataloguing in Publication Data: a catalogue record for this book is available
from the British Library.

ISBN: 978 1 84400 506 2

Printed and bound in China

dedication

This book is dedicated to our grandmothers, who are truly loved and missed. They always had the luxury of time when it came to their grandchildren. They inspired us to cook delicious food. They never worried about their waistlines (this made us worry about ours). They always had an opinion on everything: even if they were wrong, they were right. They handed down to us a wonderful sense of *Yiddishkeit*. They gave us heartfelt cuddles within their enormous, non-surgically enhanced breasts. They made our families' lives more colourful with their ability to start *broygeses*.

To Grandma Kitty and Nana Lily.
We are sure you are *shlepping naches*
from up above!

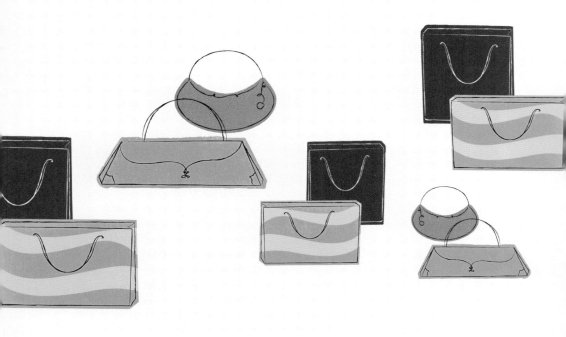

contents

acknowledgements

As I'll explain in a minute, this book grew out of a midlife crisis. I have to say that Tracey and I found this crisis great fun, complete with loads of hysterical telephone conversations and jumping up and down outside very posh offices in the West End (certainly recapturing our youth – circa age 11). Through all the excitement I actually lost loads of weight without even having to sniff the gym – bloody marvellous! So as I sit here thinking about my midlife crisis, my advice would be that if you are contemplating one of your own, and you are a Jewish Princess, just make sure it's one that is PPP: Positive, Productive, and of course, Princess-like.

Now, on to the business at hand, which is thanking people.

We are surrounded by incredible friends. I say 'incredible', not only because they seem to know everyone in the world, but also because of how unselfish and generous they have all been with their advice. It really has been a revelation that all my friends' husbands and boyfriends are actually incredibly knowledgeable, professional and *clever*, because when we go out and about, visit the latest restaurants, etc., etc., you only think of them as *Husbands* or *Boyfriends* (need I say more?). You never think of them heading up boardroom meetings or taking on their own global, midlife crises. They are just people you love and adore, take the mickey out of and have a great time with.

We would like to thank our mothers, Sandra Chester and Helen Fine (yes, you can now tell your friends); our fathers, David Chester and Tony Fine, who always answer emergency JP phone calls, even on the golf-course; and inlaws Bobbie and Irvin Tarn, whose toilet stop has proved invaluable on the way up to those West End meetings. We would, of course, like to thank Georgie's husband, RAT, and Tracey's partner, Richard de Smith, who have been our 'rocks'; our children, Cassie, Eden, Darcy, Max and Channie, whose talents have contributed to this work;

and our brothers and sisters, nieces and nephews, who have enjoyed sampling The Ultimate Friday Night Dinners again and again and again.

Then there's Anne Marie Owens (The Hairdresser), who knew Dorie Simmonds (The Agent), who knew Quadrille (The Publisher), where we met the wonderful Jane O'Shea. Thanks, too, to Stephen Marks, for all the phone calls regarding net and gross – yes, we've finally got it ('gross you get, net you vet'); Paul Taylor, for slotting us in between his high-powered business meetings; Roger Law, for meeting us for a cup of tea, a slice of cake and inspirational advice; William Miller, for your time and patience; and Ray Simone, for believing in us.

And thank you Mel Goldberg (you were there from the start); Andrew Thompson, our lawyer (who didn't have a clue what he was taking on); and Debbie Jones, our bank manager (we love you – you gave us a credit card!). We would like to thank our most understanding editor, Jamie Ambrose, who has been in intensive JP training; designer Claire Peters; PR (Princess Relations) people Clare Lattin and Emily Sanders; and all at Quadrille for making our Princess wish come true.

Finally, big thanks to the Jewish Princesses: Karen Gerrard, Elaine Grant, Zilda Collins, Lisa Marks, Sima Fine, Deborah Bright, Mandy Stanley, Auntie Rosalind Chester, Sharon Payne, Michelle Grossman and Sylvia de Smith for parting with their precious family recipes, and last but not least, Andrea (Princess G's au pair).

1

introduction

princess philosophy

I'm proud to admit that I am a Jewish Princess. I believe that every woman has a little bit of a JP in her – or if she doesn't, then she certainly should.

So just what *is* a Jewish Princess, I hear you ask? Well, the usual stereotype seems to be 'spoilt little rich girl who spends all her days shopping and beautifying herself' – let me tell you, that is simply untrue and quite unfair. After all, I never spend *all* day shopping and beautifying myself! I just like to enjoy life, and part of the enjoyment of life is food and feeding your family.

Do I love cooking? 'Not particularly' is the answer to that, so why write a cookbook? Well, as I approached my fortieth birthday, I realized that I was having a midlife crisis – but of course, being a Jewish Princess, mine was taking a rather different turn. I wouldn't contemplate an affair, and I certainly wasn't going to have a nervous breakdown or cut off all my hair (heaven forbid!). No, my crisis was going to be *productive*, and as I seem to have inherited the good-cook gene from my mother (who now no longer cooks, but can be found somewhere in a Floridian mall) and my grandmother-over-sholom (a Jewish expression for when we speak of the passed-over – not Passover: that's a Jewish holiday), a cookbook seemed like the most natural thing to do. I mean, we all have to eat, and even Jewish Princesses can't go out to eat *every* night.

And if I have to do something, I like to do it well, even if it's not exactly at the top of my 'Have to Do' list (which, by the way, is always *very* long).

So when the idea of *The Jewish Princess Cookbook* came to me, I immediately thought of my friends – after all, who can you share all the memories with? I phoned my oldest and most Princess-like friend, Tracey, to ask her what she thought of the idea and whether she wanted to be involved. She considered it for about half a second and jumped aboard.

The first thing we did was to delegate our jobs on the book: I made lunch; she made the cake.

the power of food

Food and eating lie at the very centre of Jewish culture. I really think the Old Testament was our first menu – the festivals give us such wonderful delicacies. When I think of Chanukah, I think of doughnuts. The Jewish New Year? Honey cake. And so it goes on and on.

Just think of your favourite cake. Or, if you're not a 'cake' person (*I* certainly don't know who you are, but maybe you're out there...) think of your favourite food. How does it make you feel?

When someone takes the time and trouble to make your favourite food, how do you feel about them?

Grateful. *Happy*.

Now, when people are grateful and *happy*, they are in your power. So forget the power of speech, the power of song, the power of the written word; let's get down to basics and realize that with a few eggs, some flour and a good chunk of chocolate we have the power of making others *happy*. And happiness leads to a loving home, a loving partner – and of course, lovely *things*.

It really is true that the way to a Jewish man's heart is through his stomach. If you asked Jewish men whether they would prefer a nice bowl of chicken soup or a romp upstairs, I'll bet eight out of ten of them would go for the chicken soup – especially if it contained matzo balls.

Yet the Jewish Princess of today isn't like her mother or grandmother; she's not going to be a slave to the house or the kitchen. Today's Jewish Princess wants to run a wonderful home, look good, produce lovely food, look good, look after her children – did I mention 'look good'? Sure, we want to have our cake and eat it, but we also want to look as if we haven't!

In order to do this, we have to have some tricks up our sleeves. While I have no objection to cheating in this area, and will admit sometimes to using convenience food and doctoring it to make it look like my own (oh, come *on* – we've all done it!), there truly is nothing like the satisfaction of producing wonderful homemade food.

Well, OK, there are many other satisfying things, such as 1) shopping, 2) shopping, and 3) shopping.

But I digress.

In order to have a very, very satisfying life, then, use *The Jewish Princess Cookbook* to produce food that is 1) easy, 2) non-time-consuming and 3) has fewer than ten ingredients, on average, per recipe. Oh – and of course tastes delicious and looks incredibly difficult.

If *I* can do it, *you* can do it!

So take back your take-outs, stock up your larders, forward plan, try a little bit of imagination, and use this book. You will then have the power of food in your hands.

the pay-off

The power of food produces many valuable dividends. Respect and adoration are just two of these, and when you see your friends and family loving your food, it really does give you a feeling of immense satisfaction – especially when they think you've been slaving over a hot stove, and you haven't (but no need to correct them there).

The recipes you will find in *The Jewish Princess Cookbook* are simple, easy to follow and have a fantastic success rate – but don't feel you have to be a slave to them. Whenever you cook, always think of the flavours you like and be creative. Use recipes, including the ones in this book, as guides, but if you fancy adding something or giving the recipe a twist, try it. You never know what fantastic new thing you may discover.

Always remember: it's not only the food, but how you present it that earns the seal of approval from your friends and family. So go on: take the plunge. I'm sure you will still have time left over for all those other, more satisfying, and very important things.

Which reminds me...

Got to go now – hairdresser's appointment!

jewish dietary rules
or, the kosher noshtra

Contrary to popular belief, the rules that govern the Jewish diet are *not*:

* EAT, EAT, EAT, or
* have a little (or a lot) of what you fancy, or
* when you entertain, always make sure there is enough food in case your guests bring ten friends with them.

No, the Jewish dietary rules date back to the Old Testament (a best-seller.) In this book, the Jewish people were given strict rules that applied to every aspect of eating – partly because, you know, even in those days, Jewish Princesses were thinking about that four-letter word: *diet*.

If you thought the Atkins was tricky, or found the G.I. Diet a little difficult to follow, or decided that the Blood Group Diet was far too complicated (especially if you don't know your blood group), then let me put all this in perspective. These are nothing – NOTHING – compared to following the 'K Diet', otherwise known as keeping kosher.

The Jewish kosher diet puts all others in the shade. It's complicated, you're on it for a lifetime, and sometimes it is very tempting to fall off the wagon. As far as the latter is concerned, it is up to every Jewish Princess to decide how disciplined she is prepared to be, but for anyone else who wants to try it, here are the highlights.

kosher meat

When it comes to meat, to follow the kosher diet you must only choose from animals that chew the cud and have split hooves – so beef and lamb are in and pork, camel and horse are out. Which is fine with me; I personally would never dream of eating Black Beauty (or a camel).

Not only are certain animals forbidden, but certain cuts of meat must also be avoided. However, if you buy your meat at a kosher butcher, you will never go wrong, as I can assure you they are very strict about the kinds and cuts of meat which are stocked on their shelves.

Fish are a little more straightforward. You can eat all fish which have scales and fins, but all types of shellfish are not permitted.

treif and parev

All unkosher food is known as *treif*, which actually means 'torn.' This does not mean torn between choosing Parma ham or just having the melon; it means you're not allowed to eat flesh that has been torn from one animal by another animal. All animals killed in this manner are forbidden.

Also, milk and meat must not be eaten together at the same meal, so no hamburgers – or should I say beefburgers? – with milkshakes.

Foods that contain neither milk nor meat are known as *parev*. This means these foods can be eaten with either meat or dairy meals – for example, fish, fruit, vegetables and non-animal, manufactured food products, such as non-dairy creamers.

If this all sounds like a lot of palaver, there is a bright side. For every Jewish Princess, keeping a kosher home means the added bonus of being able to indulge in buying not one, but at least *two* sets of crockery, cutlery, saucepans and linen, in order to keep all those milky and meaty dishes separate. Of course, you should never need an excuse to hit the housewares department of your favourite store, but this at least validates the urge.

As you can see, the basis of a kosher diet is something that requires discipline and willpower. Luckily for you, I've already done the hard work of making sure that all the recipes in this book are kosher.

For all Jewish Princesses who make the decision to follow this way of eating, I just want to say:

Mazel tov! You're following the oldest diet in the book.

What does a Jewish Princess
make for dinner?

Reservations!

a word about life

Now before you start trying the recipes, I want to share with you a very important part of JP philosophy. You might not catch on immediately, but bear with me, please, and read on to the end. There's plenty of time for cooking afterward.

I love Champagne, especially Princess Pink Champagne (naturally). I love the bottle: curvaceous and crowned with gold. A dry *pop!* The cork is released, and there's the slow descent of the liquid into a tall, slim glass. The colour is unique: like liquid candlelight. This perfect drink is to be sipped and savoured, and it reminds me of all the good times in my life.

Except, actually, the first time I ever drank it.

I was 16 years old, out on a date and dressed to kill in a new lilac leather dress (well, it *was* the early '80s). I wanted to impress this boy-man sitting in front of me, and being a very sophisticated teenage Princess, I ordered a Champagne cocktail. Unfortunately, nerves and an empty stomach got the better of me, and I spent most of the evening in the ladies', trying to dry my leather belt which accidentally got flushed down the loo. This unfortunate incident did not cool my ardour for Champagne, however. I just left out the cocktail part after that and went for the pink stuff straight.

And here's where the philosophy bit comes in. You see, above all else, the Jewish Princess knows that life is for living – that's with a capital 'L'. This isn't a dress rehearsal, after all (though dresses do, of course, play a very important part in every Jewish Princess's life).

The first rule of being a JP is this: if there is an excuse to celebrate – like buying your first pair of Jimmy Choos, or even when (not if) the first recipe you try in this books works a treat – then bring on and bring out the Champagne. This cool, delicious beverage makes every occasion a celebration.

You could try it right now, just to get the hang of it. Fill your glass and toast: **to life** – *lechayim!*

the princess pledge

* I pledge to have a clean and tidy home (but I never said I would do the cleaning and tidying).

* I pledge to acknowledge and embrace my *mishegasses*, as they make me who I am.

* I pledge that I will be prepared to go on holiday, even at very, very short notice.

* I pledge to buy lots of evening shoes and make sure I wear them.

* I pledge to have the bling, but not to wear everything (at once).

* I pledge that if I find an item of clothing that really suits me, I will buy it in more than one colour.

* I pledge to give rather than receive (except when it comes to jewellery).

* I pledge that I will make the most of myself, embrace the good bits and visit the plastic surgeon for the bad bits.

* I pledge that when I visit public toilets, I will do the 'Princess Perch'.

* I pledge that I will respect my manicurist, hairdresser and beautician, for they are in the background, allowing me to step into the foreground.

* I pledge to treasure my true friends, as they are the diamonds that every girl needs.

* I swear allegience to Gucci, Louis Vuitton, Prada, D&G, Dior, Missoni, DKNY, Chanel, Yves Saint Laurent, Fendi, Mui Mui, Mark Jacobs, Tod's, Jimmy Choo, Etro, Moschino, Calvin Klein, Hermès, Lanvin, Givenchy, Thierry Mugler, Roberto Cavalli, Marni, Georgio Armani, La Perla, Juicy Couture, Seven Jeans, Citizens of Humanity, Escada, Bottega Veneta, Alexander McQueen, Versace, Vivien Westwood, Diane von Furstenberg, Stella McCartney, Burberry, Emanuel Ungaro, Ralph Lauren, Chloé, Valentino, Graff, Cartier, Kauffman de Suisse, Tiffany, Asprey, Rolex, Chopard, Van Cleef & Arples, David Morris, Boodles and Bvlgari. I apologize to any I've missed, but I'm sure I'll be visiting your store soon!

2

appetizers

appetizer surprisers

I really look forward to visiting a new, trendy, 'spendy' type of restaurant. All too often, however, we end up at a 'Once-r': the restaurant you visit only once because the service and food are so bad and the prices so high that you'd never dream of going back for 'seconds'. It generally happens this way:

As I take in the beautiful surroundings, I hope the menu will arrive promptly so I can savour my glass of the pink stuff. Eventually the waiter notices me (the one at the centre table, gesticulating wildly) and we order – well, I order, as Hubby seems to think I'll somehow magically know what he wants. At this point I have to resort to sign language, either because the waiter doesn't speak a word of English, or because the restaurant is so noisy (due to designer stone floors, stainless-steel walls and chattering people) that normal speech is impossible.

Then The Wait begins – and so does my sense of Food Foreboding. I've ordered a *hot* appetizer, and seeing as the encounter with the waiter hasn't gone well, I'll be surprised if it's not lukewarm when my fork makes contact.

What is it with Once-r restaurants? They charge a fortune, yet make you wait an eternity before delivering anything. By the time the appetizer appears, you've already eaten a loaf of bread, a bowl of olives and nuts from the bar. Now, I usually find the appetizer the most enjoyable part of the meal. Why? Well, by the time I've finished the breadbasket, the olives and nuts, had words with the manager, and noshed my way through the first course, I'm simply too full and too exhausted to enjoy anything else.

After many such disastrous dinners I now really look forward to being entertained by friends or to entertaining them in my own home. Even though my abode may be a little last season, the chairs are comfy, the pink stuff is ready and waiting, and the food always arrives on time. I also know that when I serve appetizers (at the *correct* temperature), my guests will enjoy them, but they'll still have plenty of appetite left to relish the rest of their meal – and I *know* they'll be coming back for seconds!

pink grapefruit with brown sugar

serves *as many as you like!*

pink grapefruit
  (allow 1 per person)

dark brown sugar
  (2 teaspoons per person)

Peel the pink grapefruit, removing all of the pips and as much pith as possible.

Cut each into segments.

Place the segmented grapefruit into a frying pan and sprinkle with the sugar.

Cook briefly over a high heat – as the sugar melts, toss your grapefruit segments so that the fruit is well-coated. This whole process takes less than two minutes, so you could say ready in a jiffy – or should I say a Jaffa?

Serve in individual glass dishes.

This is a delicious, easy starter, warm and sweet on the outside and cool and zingy on the inside.

melon balls in champagne

serves 6

1 cantaloup melon
1 honeydew melon
about 175ml champagne, or
    other white sparkling wine

4 tablespoons runny honey
whites of 2 medium eggs, beaten
110g caster sugar

Ball both melons and mix the pieces into a bowl.

Mix the Champagne and honey, then pour over the melon balls.

Chill.

Take some Champagne glasses and dip the rim of each into the beaten egg-white, then into the caster sugar.

Spoon the melon mixture into the glasses, making sure there is plenty of liquid in each glass.

Lechayim!

watermelon with feta

serves *as many as you like!*

1 watermelon (preferably
   seedless) – allow five small
   pieces per person
vodka
feta cheese, cubed – allow
   ten cubes per person

virgin olive oil
1 basil leaf per person,
   for decoration

Cut the watermelon into slices and remove the rind (and seeds, if there are any). Place the slices in the shape of a flower on each plate.

Drizzle a little vodka over the melon.

Place ten small cubes of feta cheese around the edge of each plate.

Drizzle a small amount of olive oil over the cheese.

Place a basil leaf in the middle of each plate for decoration.

A delicious, refreshing, quick and simple appetizer that looks wonderful. Jewish Princesses will love it!

chopped herring

serves 8

500g sweet herring fillets,
  with onion
2 eating apples, peeled
  and grated
2 medium hard-boiled eggs

6 digestive biscuits
1 tablespoon ground almonds
1 slice of white bread,
  crust removed
2 teaspoons caster sugar

Drain the herrings and pat dry with a kitchen towel.

Mix all of the ingredients in a food processor until smooth.

This is delicious served with toasted black bread and cream cheese.

easy salmon pâté

serves 6 as an appetizer, 4 for a girly lunch

275g (roughly 2 fillets) red
  salmon, skinned
275g (roughly 2 fillets) rainbow
  trout, skinned
1 bunch of dill
1 medium onion, peeled
2 lemons

salt and black pepper to taste
5 teaspoons creamed horseradish
1 heaped tablespoon sour cream
3 tablespoons low-fat
  natural yoghurt
175g smoked salmon, 'un-leaved'

Wash and place the salmon and rainbow trout fillets in a saucepan along with half the dill, the onion and the juice of one of the lemons. Season with salt and black pepper, then cover with water.

Bring the fillets to the boil and boil for one minute.

Turn off the heat (if you're using an electric hob, move the fish away) and place a lid on top. The fish will carry on cooking in the hot water.

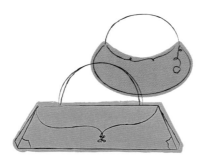

After 30 minutes, remove the fillets from the saucepan and place in a bowl.

In a separate bowl, mix the horseradish, sour cream, yoghurt and the juice of half a lemon.

Add the rest of the dill, finely chopped (I use scissors), and some salt and pepper to season.

When the fillets are cool, add the creamed horseradish mixture and blend well until the mixture forms a smooth pâté. Refrigerate.

Take a piece of foil or greaseproof paper and place the smoked salmon on top. Season with black pepper and lemon juice.

Spoon the fish pâté onto the smoked salmon, leaving a gap at the bottom so that you can easily roll the salmon lengthwise over the pâté.

Fold the smoked salmon over the pâté, making a parcel in the shape of a tube.

Wrap the foil or greaseproof paper over the salmon so that it looks like a Christmas cracker, then refrigerate.

Before serving, remove the foil or paper and cut into slices. You can decorate it with thin slices of lemon, if desired.

This starter looks very impressive indeed – so give yourself a pâté on the back!

caramelized onion tomato and brie tart

serves 6

500g ready-to-roll frozen puff
  or shortcrust pastry
a sprinkling of plain flour
1 x 300g jar caramelized onions
1-2 punnets small, firm tomatoes
  (cherry tomatoes are great for this:
  they're sweet and look the biz!)

250g mild, creamy brie
dash of olive oil
salt and pepper to taste
fresh basil leaves (enough to
  scatter over the dish)

Preheat oven to 180°C/350°F/gas mark 4.

Roll out the pastry so that it is just large enough to fit a 28cm flan tin. Lay this in the tin.

Sprinkle plain flour lightly over the surface of the pastry, then spread a thin layer of caramelized onion over the top.

Halve the tomatoes and lay them around the pastry in rings, slightly overlapping, until you get to the centre. Cover all areas quite tightly, as the tomatoes will shrink when cooked.

Slice the brie thinly and dot it over the top of the tomatoes, leaving a 2.5cm gap between pieces as it will melt and spread sufficiently.

Sprinkle with a dash of olive oil, then add the salt, pepper and basil.

Bake in the preheated oven for 35 minutes. The pastry should be golden brown.

Serve warm or cold.

This simple dish looks spectacular. It's lovely for lunch.

The first Jewish Princess, Eve,
decided that her partner needed
to be put on a healthier regime.
So for starters he would have
to eat a piece of fruit a day.

Unfortunately, she chose the
wrong one.

roasted asparagus with lemon mayo

serves 4

2 tablespoons low-fat mayonnaise
2 tablespoons sour cream
juice of 1 lemon
  (divided into two portions)

rock salt and black pepper
2 large bunches of asparagus
extra-virgin olive oil

First, make the lemon mayonnaise. Combine the mayonnaise, sour cream, half the lemon juice, and salt and black pepper to taste. Refrigerate.

Preheat the oven to 200°C/400°F/gas mark 6.

Wash and trim the asparagus spears and place them in a non-stick roasting tin.

Drizzle them with the olive oil and the other half of the lemon juice. Turn the asparagus spears over so that they have a fine coating of the olive oil and lemon juice on both sides.

Season with black pepper and rock salt to taste.

Roast the asparagus for 20 minutes, turning the spears over halfway through cooking.

Take out of the oven and leave to cool. Once cold, refrigerate or serve straight away.

To serve, place the asparagus on a serving dish with the mayo on the side.

Delicious for dunking!

roasted red and yellow peppers

serves 6

7 red peppers
7 yellow peppers
1 x 150g (drained weight) tin
pitted black olives
1 dessertspoon balsamic vinegar

olive oil to coat the peppers
(infused with garlic or basil,
if you prefer)
2 teaspoons caster sugar

Preheat the oven to 180°C/350°F/gas mark 4.

Slice the peppers into large chunks and deseed.

Rinse the olives and add them to the peppers.

Place in an ovenproof dish and pour on a good splash
of balsamic vinegar.

Add enough olive oil so that the peppers are covered.

Sprinkle the sugar over the top.

Mix the ingredients together with a spatula.

Bake in the preheated oven for one hour, turning the peppers
every 20 minutes so they do not burn, until the vegetables are soft.

*This dish can be served on its own, or for an appetizer you can
make some toasted French bread pieces and put the peppers on
top, along with a handful of basil. For a main course, just add a
handful of cherry tomatoes to the above. It also makes a delicious
pasta sauce.*

salmon fish cakes

serves 6

2 x 180g tins red salmon
  (it's easier to buy boneless)
1 large onion, grated
1 large egg, beaten

salt and black pepper to taste
50g medium matzo meal
corn oil, for frying
sea salt, for serving

Combine the salmon, grated onion and beaten egg. Add salt and pepper to taste, then add the matzo meal to bind the mixture.

Form the mixture into patties and fry in a wok or frying pan, using enough corn oil to cover. Fry until light brown.

Just before serving, grind a little sea salt over the fish cakes. They're extra-delicious if you serve them warm.

If you wish to give these a Thai-style twist, then add some chopped coriander and a little lemon-grass paste. Use kitchen paper to rest the fried fish cakes on, as this will remove any excess oil.

sesame chicken balls

serves 8 as an appetizer

500g chicken mince
4 teaspoons teriyaki marinade
3 tablespoons fine matzo meal
½ teaspoon caster sugar
2 medium eggs

half an onion, grated
salt and black pepper to taste
275g sesame seeds
vegetable oil, for frying

Mix together all the ingredients except the sesame seeds. Season well with black pepper, adding a little salt to taste.

Wet your hands and roll the mixture into little balls, then dip each into a bowl filled with the sesame seeds, covering the chicken mixture.

Fry in vegetable oil until cooked.

Serve with dipping sauce on the side (you can use a sweet chilli sauce or mango chutney), or serve with Chinese cabbage salad for a little bit of 'Ko-chin' – Kosher Chinese!

spinach bake (kugel)

serves 6 as a starter, or 4 for lunchtime with the girls

350g fine dry pasta
  (such as *lockshen* noodles)
350g spinach
25g unsalted butter

1 packet onion soup mix
350g mascarpone cheese
3 medium eggs
1 handful of pine nuts

Preheat the oven to 180°C/350°F/gas mark 4.

Boil the pasta according to instructions on the packet until it's *al dente.*

Put the spinach in a saucepan with the butter and gently fry until limp.

Combine the soup mix with the cheese and eggs. Add the cooked pasta and spinach and mix well.

Grease an ovenproof flan dish and fill with the mixture.

Sprinkle the pine nuts over the top.

Bake for an hour until crispy, or 15 minutes if cooking in individual greased ramekins.

Delicious served either hot or cold.

latkes (potato pancakes)

makes approximately 50 cocktail-sized

2.25 kilos maris piper or other
  good-quality potatoes
3 large onions
2 tablespoons self-raising flour

salt and pepper to taste
3 large eggs
corn oil, for frying

Grate the potatoes and onions. Put them in a colander and press out all the liquid: place a sheet of kitchen paper on top, then place a heavy saucepan on top of the paper and leave for five minutes.

Put the potato-onion mixture into a bowl and add the flour, seasoning and eggs.

Bind the mixture with the palms of your hands (take off all your jewellery first!). Take a loaded dessertspoon of the mixture, form it into a small round ball and flatten it by pressing gently. At the same time, squeeze out any excess liquid.

Put a generous amount of oil in a large frying pan and heat.

Place the *latkes* in the pan and semi-deep-fry them. When one side is lightly brown, turn over and cook the other side.

Remove from the oil with a slotted spoon, drain, then place on kitchen paper to absorb the oil.

Eat immediately, or freeze. If frozen, place on a baking tray lined with parchment and heat in a moderate oven for 15 minutes.

Make these latkes as small as you like for cocktail items, or medium-sized to serve with a main course. Delicious dipped in apple sauce.

3 salads

salads your personal trainer would be proud of

I first discovered my love for salad as a child in the 1970s. Thinking back, my mother was a nutritional genius, as she turned my brother and me on to raw vegetables with her unique 'Clock Salad'. This was a very '70s affair, straight out of *The Wonder Years*. It was a midweek supper which consisted of a large plate that had been sectioned with chopped vegetables, hard-boiled eggs and even grated cheese, like a pizza. We were allowed to choose our own 'sections'. Because my mother got us involved and excited about which vegetables we could choose, my love for salad was born.

Of course, such things can be taken to extremes, particularly when you're least expecting it – or should I say when you're expecting? When I was pregnant with my third child, for example, I immediately had a strong craving for raw cucumber (thank G-d, not doughnuts).* I sniffed cucumber, munched whole cucumbers down in one and even once had to leave a dinner party in search of a cucumber, smuggling it in and nibbling it out of my handbag!

Sometimes I still get a craving for salad and rush to my fridge to start chopping. Often, I'll use not only crispy, cold ingredients but I'll be inspired and add roasted vegetables, fruit or even warm meat to make my own personal Princess salad.

Salads are colourful, crunchy and, for the Jewish Princess, guilt-free eating on a plate – well, they're guilt-free as long as you don't add too many fattening ingredients. When ordering a salad in a restaurant, for instance, I always ask for the dressing on the side; this saves on those unwanted calories. So go on: dive in!

*You may be wondering why this Princess writes G-d like this. I was always taught that you should never take G-d's name in vain, even when writing. If you don't spell it out, it doesn't count...

coleslaw

serves 6

1 white cabbage, shredded
2 large carrots, grated
5 tablespoons low-fat mayonnaise

1 tablespoon double cream
salt and black pepper to taste

Mix everything together thoroughly.

If you want 'that bit extra', add one handful of flaked almonds and one handful of sultanas.

crunchy fruit and nut rice salad

serves 10

225g brown rice
225g white rice
175g dried apricots
175g dried figs
175g sultanas
salt and black pepper
110g pine nuts
110g mixed nuts
110g toasted sesame seeds

for the dressing
60ml white-wine vinegar
125ml olive oil
2 teaspoons chopped fresh
   oregano (or 1 teaspoon dried)
2 teaspoons caster sugar

First, make the dressing by combining all the dressing ingredients.

Boil the brown and white rice separately and leave to cool.

Chop all the fruits into small pieces, mix well, then add the rice and season to taste.

Stir in the dressing.

Just before serving, add the pine nuts, mixed nuts and sesame seeds and mix well.

Crunchy, colourful, and gives you loads of energy.

cucumber salad

serves 8

2 cucumbers
4 spring onions
4 tablespoons quark
   (or other low-fat soft cheese)

1 bunch of chopped chives
6 mint leaves, chopped
salt and black pepper to taste

Wash and finely slice the cucumbers and spring onions.

Add the Quark, chives and mint leaves.

Mix well and season to taste.

This is a lovely, fresh-tasting salad.

green bean and tomato salad

serves 6

400g green beans
200g small, tasty tomatoes
   (sweet cherry-type)
salt and black pepper to taste

approximately 3 tablespoons
   balsamic vinegar
1 handful fresh basil leaves

Wash and trim the beans. Place in a saucepan and cover with water.

Bring to a boil, then reduce the heat and simmer until the beans are cooked but still have a 'bite' to them. Drain and place them in a bowl to cool.

Wash the tomatoes. Fill a heatproof bowl with boiled water. Place the tomatoes in the bowl and cover for five minutes.

Remove the tomatoes and peel them – the skin should come away easily.

Chop the tomatoes and add to the beans. Season to taste with salt and black pepper.

Add a good splash of balsamic vinegar – approximately three tablespoons.

Add the chopped basil.

Refrigerate until ready to serve.

This can be served as an accompaniment to a meat or fish dish, adding colour and a crunch.

diamond carrot salad

serves 6-8

3 packets grated carrots
   (or around 1kg if you insist on
   grating your own – but I don't
   know why you would)
1 large orange, sliced

2 tablespoons lemon juice
2 tablespoons caster sugar
225ml fresh orange juice
110g sultanas

Mix together the first five ingredients, then add the sultanas.

This is even better if you let the sultanas soak overnight in the orange juice.

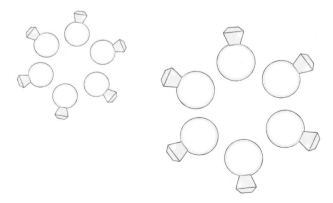

A Jewish Princess goes to visit her specialist.

She comes home and tells her husband she has been put on a very specific diet: she is only allowed carats – 2.5 and over!

oriental cabbage salad

serves 8

1 white cabbage
1 orange, cut into small pieces
1 mango, chopped
2 handfuls of raisins
125ml smooth orange juice

balsamic vinegar
olive oil
salt and black pepper
1 handful of flaked almonds

Dice the cabbage, orange and mango and mix together.

Add the raisins.

Dress with the orange juice, a good splash of balsamic vinegar and olive oil (to taste). Mix thoroughly after each addition.

Season with the salt and black pepper.

Sprinkle with the flaked almonds.

Delicious served with Sesame Chicken Balls (see page 39).

carb-loading pasta salad

serves 6-8

450g penne pasta – use brown
  pasta for a healthier dish
1 x 675g jar sweet-and-sour
  pickled cucumbers, chopped
400g tinned tuna fish

1 x 250g tin sweet corn, drained
10 tablespoons light mayonnaise
4 tablespoons tomato ketchup
salt and black pepper to taste

Cook the pasta according to the packet instructions and drain.
(Do not overcook it.)

Leave the pasta to cool.

Add the pickled cucumbers, tuna and sweet corn.

Mix the mayonnaise and ketchup together to make a Marie
Rose dressing.

Mix into the above ingredients and season to taste.

*This is very quick and easy to do. Make sure you always
keep these ingredients in your store cupboard for those
surprise lunch guests.*

chinese duck salad

serves 6

1 large duck
1 onion, peeled
1 teaspoon garlic purée
110g mixed salad leaves
1 mango, cubed
110g blanched peanuts
110g mange tout, blanched
110g bean sprouts
ready-made plum sauce
salt and black pepper to taste

for the dressing
1 teaspoon ground ginger
2 tablespoons brown sugar
1 teaspoon dijon mustard
1 tablespoon clear honey
300ml smooth orange juice

Preheat the oven to 190°C/375°F/gas mark 5.

Prick the duck all over with a fork. Put the onion and garlic purée in the duck's cavity, then place the duck on a rack over a roasting tin and cook in the preheated oven for two hours, or until the meat is cooked, turning the bird over every 30 minutes. Drain off any liquid that collects in the tin. Leave the duck to cool.

Prepare the dressing by mixing all the dressing ingredients together.

Shred the duck meat.

In a serving bowl, place the salad leaves, mango cubes, peanuts, mange tout, bean sprouts and duck meat. Pour on the dressing, toss, then season to taste.

Drizzle the plum sauce over the top just before serving.

It is so much easier to buy a ready-cooked duck!

parsley potato salad

serves 6

24 new potatoes
  (allow 4 per person)
1 bunch of flat-leaf parsley
juice of 1 lemon

virgin olive oil
125g black pitted olives, sliced
salt and black pepper to taste

Wash the potatoes and place them in a saucepan. Cover with water, bring to the boil, then turn down the temperature and simmer until the potatoes are cooked (not too soft – just so a knife will easily cut through).

Drain the potatoes and allow them to cool.

Add the chopped parsley, the lemon juice, a good drizzle of virgin olive oil, and the olives.

Season to taste.

A lovely way of making a healthier potato salad instead of a cholesterol-loaded killer!

roasted vegetable salad with goat cheese

serves 6-8

8 carrots, peeled and cut
   into batons
8 parsnips, peeled and cut
   into batons
2 sweet potatoes, peeled and
   cut into batons
2 aubergines, peeled and cut
   into batons

virgin olive oil
rock salt
1 bunch fresh coriander
1 handful of walnuts (optional)
75g soft goat cheese

Preheat the oven to 180°C/350°F/gas mark 4.

Place all the prepared vegetables in a roasting tin.

Drizzle with the olive oil.

Sprinkle with the rock salt.

Roast in the preheated oven for 45 minutes, turning the vegetables halfway through cooking.

Remove from the oven and allow to cool.

Place the vegetables into a large bowl. Add the chopped coriander and walnuts (if using).

Spoon in small balls of the goat cheese.

Drizzle with more virgin olive oil, if desired.

If you don't like goat cheese, use mozzarella balls instead.

tuscan tomato salad

serves 6

4 beef tomatoes, sliced
1 punnet small tomatoes (such as
   santa: sweet grape tomatoes),
   cut in half
6 vine tomatoes, sliced
1 red onion, finely chopped
2 bread rolls or a day-old
   baguette, cubed

1 handful of fresh basil
   leaves, chopped
4 tablespoons virgin olive oil
4 tablespoons balsamic vinegar
salt and black pepper to taste

Combine the tomatoes, onion, bread and fresh basil leaves.

Add the virgin olive oil and balsamic vinegar to taste.

Season with the salt and pepper.

*Let this marinate for at least one hour in the fridge before serving.
And do try and visit Tuscany for the real thing!*

4

soup

soup: the jewish cup of tea

In my house, we don't have teatime – we have souptime. When my children get home from school, especially in the winter, some sort of soup is always there, ready to warm and comfort them. My children know that when they open the front door, they can smell home!

Yes, soup is the perfect pep-you-up. Which makes it so very sad that so many people today never even bother to make it, thinking that it takes a zillion different ingredients and way too much time and effort. But trust me: would a true Jewish Princess include a chapter on soup in her book if that were the case? With all the fantastic vegetable and chicken stock cubes and powders* (including kosher) that are so readily available today, there's no need to spend hours making homemade stock (although you can, if you insist). Instead, you just add boiling water and get going – so perfect for Princesses!

The great thing about soup is its versatility. Whether it's served in a bowl or a mug, good soup truly is a lifesaver. Soup is the Jewish cup of tea. When you are feeling ill, cold or tired, there is nothing like a good hot bowl of homemade soup. And it doesn't just taste good – it does you good, too. In fact, chicken soup has been scientifically proven to have medicinal powers.

No wonder we've called it 'Jewish penicillin' for years!

*We JPs never use stock cubes as they don't dissolve easily. However, if this is all you have in your store cupboard, then one stock cube is roughly equal to one tablespoon of powdered stock.

bean and barley soup

serves 8

1 fowl (chicken; *see* page 194), skinned, cut into eight pieces

3 tablespoons chicken stock powder

1 x 275g tin lentils, washed and drained

2 x 450g tins butter beans, washed and drained

110g pearl barley, washed

1 large onion, peeled

4 large carrots, peeled, topped and tailed and cut into chunks

1 leek, cut into chunks

1 turnip, peeled and chopped

1 whole beef tomato (skin left on)

1 small celeriac, cubed

1 tablespoon salt

12 grinds of fresh black pepper

2.5 litres water

Place the cleaned fowl in a large saucepan and add all of the other ingredients.

Bring to the boil, then turn down the heat and let simmer with the lid on for at least three hours, or until the chicken is falling away from the bone.

Remove the tomato.

When serving, shred the chicken or remove the carcass completely (optional).

This is a meal in a bowl – and it's also low on the glycaemic index.

bloody mary borscht

serves 8

12 medium, raw beetroot,
  peeled and cubed
1 medium onion
1 medium carrot
1.2 litres hot water mixed with 2
  tablespoons of vegetable stock

3 tablespoons caster sugar
black pepper to taste
1 litre tomato juice
worcestershire sauce to taste
splash of vodka

Peel the beetroot and onion, and peel, top and tail the carrot.
Place the vegetables in a large saucepan.

Add the water and stock mixture, then the sugar, and season with
plenty of black pepper.

Bring to the boil, then turn down the heat and simmer until the
vegetables are soft. Remove the carrot and onion.

When cooled, add the tomato juice, blend well, then run
the liquid through a sieve.

Add the Worcestershire sauce to taste, then the splash of vodka.
Refrigerate until ready to serve.

Serve chilled.

*When serving this soup, add a little jug of vodka for people to
help themselves. This is a fabulous, fun soup (but watch out if
you're wearing white...).*

broccoli soup

serves 6

75g unsalted butter
2 medium onions, diced
675g broccoli, trimmed
and cut into florets
50g plain flour

1.2 litres vegetable stock (make
this with 2 heaped tablespoons
of vegetable bouillon powder)
1 pinch of nutmeg
salt and pepper to taste
300ml milk

In a large saucepan over a gentle heat, melt the butter, making sure it doesn't burn.

Sauté the chopped onions until soft. Add the broccoli and cook for five minutes, or until just tender.

Stir in the flour and cook gently for one minute. Remove from the heat and gradually stir in the stock.

Bring to the boil and then reduce the heat. Cover and simmer for 15 to 20 minutes, or until the vegetables are tender.

Add the nutmeg. Season to taste with the salt and black pepper.

Put the soup into a food processor or blender and whizz until smooth.

Reheat gently and slowly stir in the milk.

A delicious and creamy soup full of vitamins – and the children won't even realize it!

A Jewish Princess goes into a restaurant
and sees they have fish soup on the menu.
She calls the waiter over and asks:
'Does the fish soup contain shellfish?'
'Yes,' says the waiter.
She calls over another waiter and asks:
'Does the fish soup contain shellfish?'
'Yes,' says the waiter.
She calls over a third waiter and asks again:
'Does the fish soup contain shellfish?'
'No,' says the waiter.

The Jewish Princess replies: 'I'll have it!'

no shellfish fish soup

serves 8

1 onion, chopped
4 carrots, peeled, topped
  and tailed, then chopped
3 leeks, chopped
2 potatoes, peeled and cubed
2 fish stock cubes
1 teaspoon curry powder
1 teaspoon cinnamon
1 pinch of saffron, softened in
  a little boiling water

half a bottle of dry white wine
2 tablespoons brandy
1 carrot, peeled, topped,
  tailed and grated
1 leek, cut into thin batons
150ml double cream
600g mixed skinless fish fillets
  (salmon, cod and haddock),
  cut into chunks
salt and black pepper to taste

Place the onion, chopped carrots, chopped leeks and the cubed potatoes into a large saucepan.

Dissolve the fish stock cubes in 600ml of boiling water and pour over the vegetables. Add more water if the vegetables are not covered.

Add the curry powder, cinnamon and saffron.

Bring to the boil, then turn down the heat and simmer until the vegetables are soft (approximately 20 minutes).

Process the soup with a hand blender.

In a separate saucepan, pour in the wine and brandy and simmer until the liquid starts to reduce.

Add the grated carrot and the batons of leek. Cook gently until the vegetables are soft.

Add this mixture to the main soup.

Stir in the double cream and add the fish. Cook on a low heat for a further ten minutes, or until the fish is cooked and flaky.

Season to taste with salt and black pepper.

This is delicious served with French bread that has been sliced and toasted with grated Emmental cheese on top and tomato aioli, which is a garlic mayonnaise with a tomato twist.

To make the aioli, just take 4 tablespoons of mayonnaise, add a half-teaspoon of tomato paste, a half-teaspoon of garlic paste and a half-teaspoon of Dijon mustard, and voilà: it will take you straight to the South of France.

honey-roasted butternut squash soup

serves 8

2kg butternut squash, cut into
   7.5cm wedges and deseeded
3 tablespoons olive oil
garlic purée
salt and pepper to taste
4 onions, finely chopped
4 carrots, topped, tailed and
   finely chopped

2 teaspoons dried parsley
4 tablespoons vegetable bouillon
   powder dissolved in 2.4 litres
   of boiled water
1½ tablespoons runny honey

Preheat the oven to 240°C/475°F/gas mark 9.

Brush the squash wedges with a little olive oil and place them on a roasting tray. Dot the wedges with garlic purée (about one teaspoon per wedge), then season with salt and pepper and roast for 45 minutes in the oven. Cool, then scoop the flesh off the skin.

Heat the three tablespoons of olive oil in a large saucepan. Add the onions and carrots and cook gently for ten to 15 minutes, or until the vegetables are soft but not brown. Add the parsley.

Pour the stock onto the vegetables, bring to the boil, then turn down the heat and simmer for 20 minutes, or until tender.

Add the squash to the stock and veg. Simmer for five more minutes.

Blend until smooth, stir in the honey, and gently heat through. Check the seasoning and adjust if necessary.

When you cut the butternut squash, be careful, as they're very tough-skinned. You don't want to end up in casualty!

jerusalem artichoke soup

serves 8

4 tablespoons olive oil
600g red onions, chopped
600g potatoes, peeled
  and chopped
1kg jerusalem artichokes,
  peeled and chopped

2 tablespoons vegetable
  bouillon powder
1.2 litres water
salt and black pepper to taste
300ml soya milk

Heat the olive oil in a large saucepan. Add the red onions and cook, stirring constantly, until soft.

Add the potatoes and artichokes, cover, and sweat for ten minutes (this means the vegetables – not you going out for a run!).

Add the stock powder, water and seasoning. Simmer on a low heat until the vegetables are soft (approximately 30 minutes).

Liquidize the soup and stir in the soya milk.

Check the seasoning and adjust if necessary.

Don't be scared of Jerusalem artichokes. Underneath the knobbly skin of these unusual, ugly vegetables lies a true delicacy.

leek and potato soup

serves 8

1 1/2 tablespoons olive oil
2 onions, finely chopped
7 large leeks, washed, peeled
  and finely chopped
4 large potatoes, cut into
  1cm chunks
2 level tablespoons vegetable
  bouillon powder

2.4 litres water
sea salt and black pepper to taste
1 teaspoon ground nutmeg
300ml double cream
2 tablespoons fresh
  chives, chopped

In a large saucepan, heat the olive oil. Add the onions and leeks and gently sauté until soft.

Add the potatoes, the vegetable bouillon powder, water, salt and pepper. Bring to the boil, turn down the heat and simmer, covered, for 20 to 25 minutes, or until the vegetables are soft.

Stir in the nutmeg.

Remove from the heat and allow to cool.

Liquidize thoroughly and stir in the cream.

Check the seasoning and adjust if necessary.

To serve, stir in a little swirl of cream in each bowl and garnish with chopped chives.

This can be made the day before to allow the flavours to infuse – and to give Jewish Princesses a little extra shopping time.

mushy pea soup

serves 8

1.8kg frozen petits pois or
  other green peas
1 large onion, chopped
3 leeks, sliced
1 tablespoon dried parsley

chicken or vegetable stock
  powder – approximately
  3 tablespoons
salt and black pepper to taste
mint leaves, for decoration

In a large saucepan, place the petits pois, onion and leeks.
Add the parsley and stock powder, fill the saucepan with enough
water to cover the vegetables, then bring to the boil. Turn down
the heat and let simmer, covered, until the peas are very soft.

Using a hand blender, liquidize the soup. Add the salt and
pepper to taste.

When the soup has cooled, push it through a sieve – this is
annoying, but it makes the soup taste incredible! To serve, put
a few mint leaves in each bowl for decoration.

*To make a fun dinner, ask your husband to go get some fish and
chips – so now you have your fish, chips and peas!*

onion soup

serves 6

5 teaspoons olive oil
4 medium onions, sliced
4 teaspoons caster sugar
1.7 litres water

3 tablespoons vegetable
  bouillon powder
2 teaspoons marmite
2 tablespoons sherry
salt and black pepper to taste

In a large saucepan, heat the olive oil, then add the onions and sweat them until soft.

Add the rest of the ingredients.

Bring to the boil, then lower the heat and cover. Leave the soup to simmer on a very low heat for about 30 minutes.

Season to taste.

When serving, toast thick slices of French bread and melt grated Cheddar or Emmental cheese on them. Chop them up and place them as croûtons in the middle of the soup.

parsnip and apple soup

serves 6

1 onion, diced
450g parsnips, washed
  and cubed
1.5 litres boiled water

3 tablespoons vegetable
  bouillon powder
5 eating apples, peeled and sliced
250ml double cream

Put the onions, parsnips, water and bouillon powder in a large saucepan. Bring to the boil, lower the heat, cover and simmer for 20 minutes, or until the parsnips are soft.

Meanwhile, put the peeled apples in a separate microwave-able bowl, cover with two tablespoons of water and cook on high for two minutes. When you take the apples out of the microwave, check that they are soft.

Add the apples to the vegetables.

Liquidize the soup and stir in the cream to serve.

This one will keep your dinner guests guessing, trying to work out all the ingredients.

tomato soup with basil-infused croûtons

serves 6

2 tablespoons olive oil
4 shallots, peeled and chopped
3 celery sticks, washed
  and chopped
2 carrots, peeled, topped,
  tailed and chopped
1 large handful of torn
  basil leaves
1 large bunch of sage leaves
2 bay leaves
3 x 400g tins chopped
  tomatoes in their juice

1.2 litres water mixed with
  2 tablespoons of vegetable
  bouillon powder
3.5 tablespoons caster sugar
salt and black pepper to taste
splash of olive oil

for the croûtons
day-old French bread
1 garlic clove
6 tablespoons olive oil
1 handful of torn basil leaves

First, make the croûtons.

Preheat the oven to 140°C/275°F/gas mark 1.

Slice the French bread into thin slices. Rub both sides with garlic.

In a bowl using a hand blender, blend together the olive oil and basil.

Brush onto one side of the bread slices.

Place in the preheated oven for 20 minutes, turning halfway through, until the bread has dried out.

Place in an airtight container.

To make the soup, heat the olive oil in a large saucepan. Add the shallots and sauté.

As they become translucent, add the celery, carrots, basil, sage and bay leaves.

When the vegetables are soft, add all the other ingredients.

Reduce the heat, cover and simmer on a low heat for a good hour or so.

Liquidize the soup.

To make it extra-special, pour it through a sieve before serving.

Serve with basil-infused croûtons.

Tomato soup is hard to get right, but this is JP perfect! Serve with croûtons and a basil leaf to decorate.

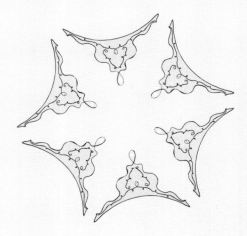

diaspora dishes:
london to new york

And now, an aside. While you recover from all those luscious soups, salads and appetizers, it's time for a little break before diving into the main courses. So put your feet up for a minute and relax while reading the following. It's educational (well, a *bissel*...).

Maybe it's because of the Jewish people's history of dispersal, traditionally known as the Diaspora, but whatever the reason, the modern Jewish Princess still loves to roam – or should I say Rome, Paris, Milan, London and New York? Cities like these, after all, are centres of cultural importance, home to resplendent art galleries, shops, magnificent museums, shops, incredible restaurants – and did I mention the most amazing shops?

To a JP, this intoxicating combination is always a recipe for success. Whenever I return from any holiday, I'm always motivated to try out in my own kitchen all the foods I have eaten and loved abroad. Almost as soon as the tyres hit the runway, I find myself running to my nearest supermarket to stock up on new foods I have experienced (OK, so I'm in the shops again), simply to bring back the flavour of wherever it is I've just been.

Throughout history, Jewish Princesses have loved to travel, and our food contains tastes and flavours that come from all over the world. This makes Jewish cooking terrifically exciting, as this hotpot of global flavours produces the most extraordinary dishes. From Spain to South Africa, London to Lithuania, Ashkenazi and Sephardi Jews (*see* Yiddish/English Glossary, page 210) have concocted a unique style of cooking that takes a little of this and a little of that to create food which family and friends adore.

So next time you come home from holiday, if you can't fit the sun into your luggage due to the other purchases you made, just remember all those flavours you tasted and savoured. With a bit of culinary creativity, you can be transported back to that special place again and again.

5

meat

meat: not always a 'rare' kosher treat

No matter how far she may wander or roam, ultimately a Jewish Princess will have to make some key decisions.

The first is: 'Do I keep a kosher home?'
The second: 'Do I keep kosher when I go out to eat?'
The third: 'Do I eat only in kosher restaurants, or eat only certain fish or vegetarian dishes and go to non-kosher restaurants?'

Everyone finds her own level. Often it relates to how you were brought up or how guilty (ever heard of Jewish guilt?) you might feel if your fork touched a forbidden food. For a Jewish Princess, it is up to her to find which foods may turn her into a frog.

I made my decision and decided to keep a kosher home. Therefore the relationship I have with my butcher is of paramount importance. Not, obviously, as important as the relationship I have with my hairdresser, but close.

I have spent many years searching for an excellent butcher. This search was a bit like my dating years. I have 'courted' a few butchers in my time, but I had to chuck them for various reasons: sending me a hairy chicken; not saving me eggs for the chicken soup (a great kosher delicacy – these golden eggs are in short supply and I am not prepared to do my butcher any favours to get them). A tough piece of beef was definitely a chucking-out offence, but the worst crime of all was when a butcher completely ripped me off for a crown of lamb by charging me an exorbitant price.

I am afraid there was A Scene.

Finally, however, after much cruising and shmoozing, I found the perfect butcher. Of course, we are on first-name terms; I know all about his family, (well, most of them work in the shop). My chickens are hairless, my beef is always tender and delicious, and he always saves me the precious golden eggs for my chicken soup – *without* any favours!

Why am I telling you all this? Simple: my friends and family are Carnivores, with a capital 'C'. We just *love* meat.

Also, many of my friends have made the decision to keep kosher when dining out, and they will eat only fish and vegetarian dishes in non-kosher restaurants. They spend their evenings eyeing up pieces of beef and chicken that come flying out of the kitchen and past them, carried by deaf waiters (Don't you find that this is a common affliction affecting most waiters? It seems to be a requirement for the job...). So when my kosher-keeping friends come to my home for dinner, it is a really special treat to tuck into meat – and there is nothing more satisfying than to take out of the oven a juicy joint and know that my guests will truly appreciate it.

Now I know that, like fine wine, there is a certain amount of snobbery regarding the cooking of meat, but I have to say, give the people what they want. If they don't like it rare now, they never will. And if they love it well done, then you should serve it that way.

It is the *enjoyment* of the meat that is everything.

So when I have guests, there are no deaf waiters (I do occasionally have a waitress, but that is only for very large gatherings), the meat is always cooked to my friends' requirements, and when I see an empty plate I know that I, for one, will get a big 'well done'.

A Jewish Princess went on a
walking holiday. She power-walked
from Egypt to Israel.

It took her forty years, but boy,
did she have great legs by the end!

chicken curry in a hurry

serves 2

3 tablespoons olive oil
2 onions, diced
2 chicken breast fillets, diced
1 x 175g tin tomatoes in
tomato juice
3 teaspoons medium
curry powder

50g sultanas
1 banana, sliced
1 x 175ml tin creamed coconut
1 teaspoon caster sugar
salt and pepper to taste
mango chutney to serve

Heat a wok or large, heavy frying pan.

Add the olive oil.

Add the chopped onions and sauté until brown, then add the diced chicken fillets and stir until cooked.

Add the tomatoes and curry powder to taste, then stir in the sultanas and the sliced banana and cook for approximately three minutes.

Add the creamed coconut and the caster sugar and continue cooking until the liquid has reduced and a creamy sauce is left (approximately ten minutes).

Season to taste with salt and pepper.

Serve with cooked rice and mango chutney on the side for a hot, romantic dinner for two.

chicken schnitzel in wine and spritely lemonade

serves 10

10 chicken fillets
3 large eggs, beaten
breadcrumbs or matzo meal
olive oil, for frying

for the marinade
12 tablespoons white wine
8 tablespoons sprite
4 garlic cloves, crushed
2 tablespoons lemon juice
1 x 15g packet of fresh chives

Mix all the marinade ingredients together and pour over the chicken fillets. Leave to marinate for a couple of hours – or even better, overnight.

Dip the chicken fillets into the beaten egg, then in the breadcrumbs or matzo meal.

Fry in the olive oil until golden brown and serve.

This unusual concoction always goes down well, even with difficult eaters. Try it and see!

gedempte chicken

serves 6

3 tablespoons corn oil
1 red onion, diced
1 white onion, diced
1 chicken, cut into 8 pieces
enough plain flour to coat the
  chicken (approximately 150g)
500ml water

1 x 500g tin passata
2 garlic cloves, pressed
2 bouquets garni
salt and black pepper to taste
6 large potatoes, cut into quarters
  and parboiled

Heat the oil and fry the onions in a large wok until they are as dark as possible without burning them.

Dip the chicken pieces in the plain flour and fry them with the onions until the chicken is slightly brown.

Transfer the chicken and onions into a deep saucepan and add the water, passata, garlic, the bouquets garni, salt and pepper.

Bring to the boil, then turn down the heat, cover and let it simmer very gently until the chicken is soft and dark-brown in colour. This will take at least one and a half hours.

Halfway through the cooking, add the partly boiled potatoes, which will continue to cook in the meat juices and will turn a dark brown.

Whenever I make this, I feel like my grandmother-over-sholom is with us at the table, enjoying the fact that all the family are together.

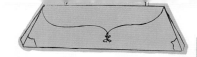

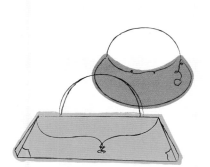

A Jewish Princess goes into her
butcher's to pick up her chickens.
'Are they in yet?' She enquires.

'No,' says the butcher. 'They'll
be back from the dry cleaners in
two minutes!'

moroccan chicken

serves 6

vegetable oil, enough to fry
    (approximately 4 tablespoons)
1 tablespoon tomato purée
2 teaspoons garlic purée
2 teaspoons cumin
1 teaspoon cinnamon
1 chicken cut, into 8 pieces
4 medium onions, peeled and
    sliced (use frozen if you like)

18 dried apricots
3 preserved lemons
    (found in most supermarkets)
300ml water
1 tablespoon chicken stock
    powder or 1 chicken stock cube
salt and black pepper to taste

Preheat the oven to 180°C/350°F/gas mark 4.

Heat the vegetable oil, then add the tomato purée, garlic purée, cumin and cinnamon.

Quickly fry each piece of chicken.

Place the chicken pieces in a roasting dish.

Fry the onions in the remaining oil until they are soft. Do not burn!

Pour the onions over the chicken.

Add the dried apricots, preserved lemons, and water mixed with the chicken stock powder (or cube) to the bottom of a roasting tin. Season with the salt and pepper.

Roast in the preheated oven for at least one hour, or until the juices run clear.

Lovely served with couscous. Just don't do those belly-dancing exercises straight afterwards...

poussin with apple and sage

serves 6

3 poussin (allow at least half a
    poussin per person)
3 gala (or similar) apples, halved
6 fresh sage leaves per poussin
shallots, peeled
carrots, topped, tailed, peeled
    and cut into chunks

chicken stock, approximately 500ml
salt and black pepper to taste
honey
olive oil
fresh oregano leaves

Preheat the oven to 180°C/350°F/gas mark 4.

Clean each poussin.

Stuff each one with two apple halves and six sage leaves.

In the bottom of a roasting tin, place the peeled whole shallots
and peeled chunks of carrots (as many as you want).

Place the poussins in the roasting dish and pour in enough
chicken stock so that the dish is at least half-full.

Season the poussin with salt and black pepper. Rub a little honey
and olive oil into the skin of each poussin.

Sprinkle fresh oregano leaves over the top.

Roast in the preheated oven for 45 to 60 minutes, or until the skin
is golden.

*This is easy to serve, and the result looks like a celebrity chef has
invaded your kitchen.*

cumberland cutlets

serves 4 to 6 *(depending on how many men you're serving!)*

8 lamb cutlets
salt and pepper to season
2 medium eggs, beaten
fine matzo meal or
    fine breadcrumbs

for the sauce
450ml chicken stock
9 rounded tablespoons redcurrant
    (or cranberry) jelly
juice of 4 oranges
1 orange, sliced, for garnish
5 teaspoons lemon juice

Preheat the oven to 200°C/400°F/gas mark 6.

Trim off most of the fat from the lamb cutlets, then sprinkle with salt and pepper.

Dip the cutlets into the beaten egg, then into the breadcrumbs or matzo meal. Put in a baking tin side by side.

To make the sauce, heat all the ingredients together until smooth and pour half over the cutlets. Cover with foil.

Cook in the preheated oven for one hour, then add the remaining sauce. Turn the oven down to 180°C/350°F/gas mark 4 and cook for a further 20 minutes.

Garnish with the orange segments.

Not, sadly, named after the Cumberland Hotel in Bournemouth, a kosher hotel that existed in the 1970s which was renowned for its copious amounts of food.

sticky lamb chops

serves 6

6 tablespoons clear honey
6 tablespoons tomato ketchup
6 tablespoons dark soy sauce

6 tablespoons water
12 medium lamb chops

Combine the honey, ketchup, soy sauce and water in a bowl.

Arrange the lamb chops lying side by side in an ovenproof dish.

Pour the sauce over the meat and place in the fridge to marinate for half an hour (or longer if desired).

When ready to cook, bake uncovered in an oven preheated to 160°C/325°F/gas mark 3 for at least an hour.

Check that the cutlets are not drying out, and turn throughout cooking, basting with the marinade.

Junior JPs' favourite.

shepherd's pie

serves 6

1 onion, diced
10g non-dairy margarine
200g mushrooms,
    peeled and sliced
450g minced lamb
1 x 400g tin chopped tomatoes
6 bay leaves
salt and black pepper to taste

for the topping
1kg sweet potatoes,
    peeled and diced
30g non-dairy margarine
1 tablespoon tomato purée
½ teaspoon cumin
½ teaspoon nutmeg
salt and black pepper to taste

Preheat the oven to 190°C/375°F/gas mark 5.

Fry the onion in the margarine until slightly brown.

Add the mushrooms and continue to fry until soft.

Add the meat, the tinned tomatoes and the bay leaves and simmer on a low heat for one hour. Don't forget to season with salt and black pepper at this point!

Transfer the lamb to an ovenproof dish.

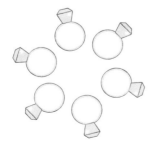

Next, make the topping. Put the sweet potatoes in a saucepan with just enough water to cover, then boil them until soft.

Drain off the water and leave the potatoes to cool.

Add the rest of the topping ingredients to the potatoes and mash thoroughly. Spread the topping over the cooked lamb.

Bake the lamb in the preheated oven for approximately 25 minutes, or until the topping has become crispy.

When you make this dish you won't need a shepherd to get everyone to the table!

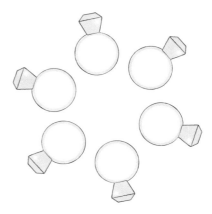

brisket with prunes

serves 6

1.5kg brisket (unpickled)
2 parsnips
6 carrots
2 onions

black pepper to taste
1 bay leaf
12 dried pitted prunes
50g onion soup mix

Preheat the oven to 150°C/300°F/gas mark 2.

Wash the brisket and place the meat in a roasting tin filled halfway with water.

Peel all the vegetables and leave them whole. Season well with black pepper and add them to the roasting tin.

Add the rest of the ingredients.

Cover with foil and cook in the preheated oven for three hours, turning the brisket every hour, until the meat is soft.

A great dish for impressing your guests, especially when served with couscous. As a bonus, you can leave it in the oven and have a little 'me time' (three hours' worth!) before they arrive.

you say corned beef, i say salt beef

serves 8

2kg centre pickled brisket
1 onion, washed but not peeled
6 peppercorns

1 bay leaf, split either end
to infuse
1 teaspoon caster sugar

Place the washed salt beef in a large saucepan and cover with water. Cover with the pan lid and bring to the boil.

When scum appears, throw away the first water and refill with fresh, clean cold water – enough to cover the beef.

Add the onion, peppercorns, bay leaf and sugar and bring to the boil.

When scum reappears, skim this away with a large spoon. Place the lid half on the saucepan and continue to simmer for approximately two and a half hours, or until the beef is tender.

Add boiling water to the saucepan if the liquid falls below the line of the beef.

To see if the beef has cooked, test it by piercing with a fork. If it comes out easily, it's ready.

Either serve straight away or allow the beef to get cold.

When cold, slice thinly and reheat in the microwave with a little of the liquor on top.

For a great sandwich, deli-style, serve salt beef on rye bread with English mustard and pickles. Salt beef and latkes (see page 41) are the Jewish 'eggs and bacon'.

cholent
(meat stew with pulses and potatoes)

serves 8

2kg stewing beef (I use top rib)
6 onions, sliced
175g pearl barley
6 large potatoes, cut into quarters
2 bay leaves
235g (drained weight) tinned
   cannellini beans
235g (drained weight) tinned
   butter beans

235g (drained weight) tinned
   borlotti beans
350g passata (or a large can of
   chopped tomatoes)
plenty of black pepper
2 teaspoon salt
2 teaspoons dark brown sugar
3 carrots, peeled, topped
   and tailed
2 tablespoons beef stock powder
   or 2 beef stock cubes

Preheat the oven to 200°C/400°F/gas mark 6.

Place all the ingredients in a large, heavy, ovenproof pot or dish (I use a large saucepan) with the beef in the centre and the potatoes around the beef.

Pour in enough water to cover all the ingredients, then cover and cook in the preheated oven for one hour.

Turn the heat down to 150°C/300°F/gas mark 2 and leave in the oven for the next seven hours.

If you're going to leave the dish to cook longer, just turn the heat down even more.

I know this breaks the ten-ingredient rule, but it's allowed due to the fact that it is simply a matter of opening cans (just watch your fingernails, of course!).

This wonderful stew is traditionally made on Friday before the Sabbath, and left in the oven overnight, ready for lunch the next day. Absolutely delicious.

tzimmes
(braised beef and carrots)

serves *as many as you like!*

1kg carrots, peeled and cubed
  (or buy prepared)
3 tablespoons soft
  light brown sugar
beef brisket or rolled rib (any size,
  depending on how many you
  want to feed)

salt and pepper to taste
olive oil, for searing
4 tablespoons cornflour
1kg potatoes, peeled and
  sliced thickly
1 x 450g tin golden syrup

Preheat the oven to 140°C/275°F/gas mark 1.

Boil the carrots until *al dente* in water to which one tablespoon of brown sugar has been added.

Season the meat with salt and pepper.

Coat a large frying pan with a little oil, heat it, and sear the meat.

In a bowl, put two tablespoons of brown sugar and two tablespoons of cornflour. Stir them together.

Drain the carrots, reserving the liquid.

Line a large, deep, heatproof casserole with a layer of carrots. Sprinkle them with the sugar and cornflour mixture.

Place the seared meat on top of the carrots in the middle of the casserole.

Sprinkle two tablespoons of cornflour over the meat.

Continue layering with carrots and potatoes, seasoning each layer as you go.

Pour the reserved carrot liquor into the casserole, making sure the meat is covered. Add water if needed.

Pour in the golden syrup, then cover the dish tightly with aluminium foil to keep it from drying out.

Cook in the preheated oven for at least eight hours. Keep checking to see if it needs some more water.

This dish is best made the day before so that you can leave it to cook overnight on a low heat. It's great to serve in front of the fire on a cold, dark evening with a big glass of red wine.

ye olde steak pie

serves 6

1kg diced steak
1 punnet button mushrooms
25g onion soup mix
150ml kiddush wine
   (or you can use
   your own red plonk)
150ml water
1 garlic clove
6 shallots, peeled
   and left whole
4 small (or 2 large) carrots,
   topped, tailed and chopped

1 x large 400g tin
   chopped tomatoes
1 teaspoon caster sugar

for the pastry
500g puff pastry
1 medium egg, lightly whisked
black pepper to taste

Preheat the oven to 120°C/250°F/gas mark ½.

Place all the ingredients in a round casserole dish.

Cover with a lid or foil, and cook slowly in the oven for at least three hours, or until the meat is soft.

Leave to cool. Check the seasoning and adjust if necessary.

Roll out the pastry (unless you buy pastry already rolled – 'Great move!' I say).

Paint beaten egg around the rim of the casserole. Add the pastry and press down the sides to seal.

Brush beaten egg over the top and make a cross in the centre with your knife (this stops the pastry going soggy).

Preheat the oven to 220°C/425°F/gas mark 7 and bake for five minutes, then reduce the heat to 190°C/375°F/gas mark 5 and bake for a further 30 minutes, until the gravy is bubbling up.

If you're concerned about the pastry (especially if you're watching your carbs), just make the casserole on its own.

fish

f.i.s.h. = finally i saved him

After Hubby and I had been going out for a few weeks, I was sure he would be ready to propose once he had been to our house for a Friday night dinner.

This, of course, led to the inevitable problem of having to consider that maybe it wasn't *me* he fell in love with, but the meal my mother served him...

But anyway.

The reason I felt I was on to a sure thing was simple. Because his mother, for some very bizarre reason, served fish on Friday nights (maybe she thought they were Catholic?), Hubby had been deprived of the joys of chicken soup, chopped liver and roast chicken all his life (*see* 'The Ultimate Friday Night Dinner', page 187). So when he arrived at my house and saw the table, when he was served piping-hot chicken soup and all the rest, he saw in my mother the promise of the woman I would become.

That night, instead of eating it, Hubby-To-Be *turned into* fish – he had been well and truly caught!

Due to the fish that was forced upon him, my lovely husband suffered for many years with a fish phobia. After all, if *you* had been given fish that looked like it had just escaped from a hospital ward (his mother wasn't the most imaginative cook), you, too, would find yourself fearing fish.

So Hubby hated fish with a passion, and whenever he had to eat at his mother's after we were married, she just served him a hard-boiled egg, as, by this time, she had finally given up. (Being a true Jewish Princess, I always politely ate my hospital fare.)

Because of his fish fear, I decided that fish should not be forced back into my husband's life. However, I didn't want to deprive myself or my children of that excellent brain food. After all, they would have school entrance exams to sit, so they would need every little bit of help they could get (including the odd hour or so of private tuition – competition is very high).

So with a few clever recipes, I was sure that my beloved could grow to have a fondness for fish. Slowly, over the years, and with the aim of preserving my husband's heart muscle, I have managed – with patience, coaxing and little phrases such as, 'Go on: just one little bite' (well, you know all men are little boys at heart) – to persuade him to eat fish.

And yes, you better believe it: now he actually *likes* it!

It really took some Princess perseverance, but after all, he *was* my soulmate. I knew that, with the right treatment, he could find a plaice – sorry, make that *place* – in his heart for fish.

So if you have a member of your family who says 'I hate x,y and z', don't give up. After all, sushi at first sight takes some getting used to; now people are in love with it the world over.

I mean, whoever thought *my* Prince would actually *challish* for a piece of raw salmon?

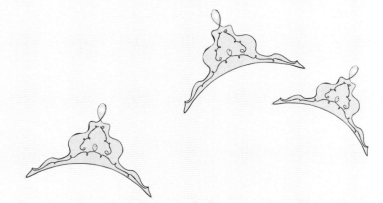

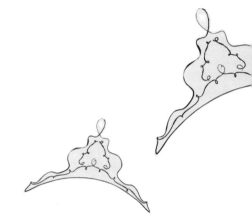

What's a Jewish Princess's
favourite wine?

I want to go to Marbella!

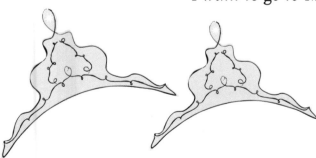

fish tempura

serves 4

2 skinless salmon fillets
2 skinless cod fillets
salt and black pepper to taste
150g tempura batter mix
(this can be purchased at
most large supermarkets or
a japanese food store)

vegetable oil
16 baby corn pieces or
baby carrots
soy sauce

Cut each piece of salmon and cod into four strips.

Season the fish with salt and black pepper.

Mix up the tempura batter.

Coat the fish with tempura batter and fry (making sure there is enough oil to cover the fish) until golden brown.

Coat the vegetables in the batter and fry until golden brown.

Place all the fish and vegetables on kitchen paper to soak up any excess oil.

Serve with soy sauce on the side to dip.

If you have chopsticks in the house, use these instead of knives and forks – it's much more fun!

haddock kedgeree

serves 4

2 large skinless fillets natural
  smoked haddock
1 tablespoon vegetable
  bouillon powder
350g rice
110g unsalted butter

1 teaspoon curry powder
1 handful of chopped parsley
3 medium hard-boiled
  eggs, chopped
425ml double cream
salt and black pepper to taste

Place the haddock into a large saucepan and cover with water.

Add the vegetable bouillon powder, and cook over a low heat until the haddock is poached.

Remove the fish and leave to cool. Retain the fish stock, but set aside one ladle of it for later.

Add the rice to the fish stock and cook, adding more water if necessary. Set aside when cooked.

Turn on a medium heat and drop half of the butter into another saucepan. Add the curry powder, chunked haddock, and parsley and sauté gently, stirring constantly.

Add half the cooked rice and keep stirring. Add the remainder of the rice and the rest of the butter. Keep stirring.

Add the chopped egg, the ladle of fish stock and the cream.

Season to taste with black pepper and serve immediately.

Haddock kedgeree is a bit like marmite – you love it or you hate it. So be careful who you serve it to!

hot and spicy fish

serves 6

for the fish balls
2 large eggs
1 large potato (approximately
   225g), peeled and boiled
   until soft
1 tablespoon dried parsley
1 teaspoon paprika
1 teaspoon ground nutmeg
1 red onion, diced
1 teaspoon garlic purée
500g white fish, minced
salt and pepper to taste

for the sauce
2 x 400g tins chopped tomatoes
55g chopped fresh basil
1 tablespoon tomato purée
2 teaspoons curry powder
2 teaspoons caster sugar
salt and pepper to taste
3 tablespoons single cream

corn oil, for frying

Place all the fish-ball ingredients in a food processor and process until everything is well-mixed.

Let the mixture rest for 15 minutes.

To make the fish balls, wet your hands with water, then take a heaped teaspoon of the mixture and shape it into an oval ball.

In a frying pan, fry the fish balls in just enough corn oil to cover until they are a golden colour.

Drain on kitchen paper to absorb any excess oil, then place in an ovenproof dish.

To make the sauce, put all the sauce ingredients into a food processor and mix thoroughly.

Preheat the oven to 180°C/350°F/gas mark 4.

Pour the sauce over the fish balls and bake in the preheated oven for 40 minutes.

These are great without the sauce to serve with drinks. Or instead of serving them as a main course, serve a smaller amount as a forspeise (see Yiddish/English Glossary, page 211).

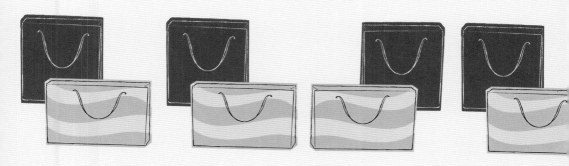

individual salmon en croute

serves 4

150ml double cream
1 x 250g tin asparagus
1 bunch watercress, chopped
1 bunch of dill, chopped
1 pinch of nutmeg

salt and black pepper to taste
500g ready-made puff pastry
4 skinless salmon fillets
1 medium egg, beaten

Preheat the oven to 220°C/425°F/gas mark 7.

In a large bowl, add the cream, asparagus, chopped watercress, chopped dill, nutmeg, salt and pepper (season well). Mix with a hand blender.

Roll out the pastry to about half a centimetre thickness.

Season the salmon with a little salt, then place it on the pastry. Cut a square large enough to cover the salmon and make a parcel.

Spoon the asparagus mixture on top of the salmon and close the pastry to make the parcel. Press down the edges.

Brush on the beaten egg and score the top of the pastry.

Place on greaseproof paper.

Bake in the preheated oven for ten to 15 minutes.

If you do not wish to use individual fillets, just use a larger piece of salmon, or make smaller ones to serve as canapés. Once the salmon en croute is cooked, slice it into pieces to serve.

lemon fish cakes

serves 6

2 medium eggs
4 slices of white bread,
  without crusts
1 tablespoon dried parsley
1 teaspoon hot paprika
1 teaspoon ground nutmeg
1 garlic clove, chopped

1 tablespoon wholegrain mustard
1 tablespoon lemon juice
500g white fish, minced
salt and black pepper

corn oil, for frying
mango chutney, for serving

Put all the ingredients, except the oil and chutney, into a food processor and process until they are all well-mixed.

Let the mixture rest for 15 minutes.

Using a heaped teaspoon, drop a spoonful of the mixture at a time into enough oil to cover the cakes and fry until they are golden in colour.

Serve on a bed of basmati rice with mango chutney.

paella marbella

serves 6

2 tablespoons vegetable oil,
  for frying
1 green chilli, deseeded
  and diced
1 medium red pepper, deseeded
  and chopped
2 teaspoons garlic purée
2 large beef tomatoes, sliced
  and cubed
1.25 litres vegetable or fish stock
2 generous pinches of saffron

200g fresh salmon
300g fresh cod
350g rice
150g fresh tuna
115g black olives, pitted
  and drained
salt and black pepper to taste
225g frozen peas
1 red onion, peeled and sliced
  into rings

Coat a large, deep frying pan with the vegetable oil (I usually use a wok), then add the chilli, pepper and garlic and cook for five minutes over a moderate heat, stirring the ingredients continuously.

Next, add the tomatoes and continue stirring for another five minutes.

Pour in the stock along with the saffron and simmer for ten minutes, keeping an eye on the mixture by stirring occasionally.

Add all the fish except for the tuna and simmer for another ten minutes.

Add the rice to the pan, mix it in with the contents and simmer for another ten minutes.

Add the tuna, salt and pepper and olives and cook for a further five minutes.

Add the peas and cook for a further five minutes.

In a separate frying pan, sauté the onion rings in a little olive oil and add on top of the paella to garnish.

So I know this recipe breaks the ten-ingredient rule, but even JPs know that rules are meant to be broken. This dish is so easy to make and looks FABULOUS! Mucho delicious!

sole colbert

serves 2

4 lemon sole fillets
salt and black pepper
flour, for dusting
2 bananas, sliced horizontally

110g unsalted butter
1 tablespoon vegetable oil
mango chutney, to serve

Wash and dry the sole fillets.

Season with salt and black pepper.

Put the flour on a plate and lightly dip the sole fillets into the flour, then pat off any surplus.

Dip the bananas into the flour, also patting off any surplus.

Heat a frying pan and melt the butter with the oil.

Fry the sole fillets gently, making sure that both sides are brown.

Do the same with the bananas.

Arrange on a plate, with the mango chutney on the side.

Husbands, boyfriends and lovers will adore this dish. Just don't invite them all over at the same time...

A Jewish Princess
went fishing.

She came home with
a new place!

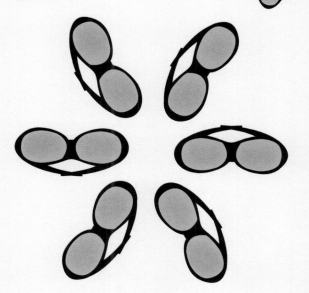

sophisticated fish pie

serves 8

2 onions, diced
2-3 tablespoons vegetable oil
2 aubergines, cubed
4 carrots, topped, tailed
   and cubed
half a cauliflower, cut into florets
2 x 400g tins chopped tomatoes
200ml double cream
3 teaspoons curry powder
1 teaspoon ground ginger
6 teaspoons caster sugar
salt and black pepper
300g smoked skinless haddock
   fillets, cubed

300g skinless haddock
   fillets, cubed
500g skinless salmon
   fillets, cubed

for the topping
5 medium potatoes, peeled
   and cubed
4 sweet potatoes, peeled
   and cubed
125ml double cream
110g unsalted butter
1 teaspoon table salt

Sauté the onions in the vegetable oil in a large saucepan. When they begin to brown, add the aubergines and carrots, and a few minutes later add the cauliflower florets.

When the vegetables have softened, add the tinned tomatoes, double cream, spices, sugar, salt and pepper. Simmer for five minutes.

Add the fish and cook for ten minutes.

Next, make the topping. In a separate saucepan, cover the potatoes and sweet potatoes with water, bring to the boil, then reduce the heat and simmer until the vegetables are very soft.

Remove from the heat and drain.

Preheat the oven to 170°C/325°F/gas mark 3.

Add the cream, butter and salt, then mash well.

Place the fish mixture in an ovenproof dish (I use a casserole saucepan, which I can just transfer from hob to oven).

Add the topping and cook in the preheated oven for 15 to 20 minutes, until bubbling.

A Jewish Princess alternative to oysters and Champagne.

sweet and sour halibut

serves 4

4 halibut fillets
1 onion
3 bay leaves
salt and pepper

4 medium eggs
2 lemons squeezed
5 dessertspoons caster sugar
cucumber, for decoration

Place the fish in a saucepan along with the onion, bay leaves, salt and pepper. Cover with water and bring to the boil.

Simmer for 15 minutes.

When the fish is flaky, take 225ml of the stock out and strain into a bowl. Allow to cool.

Carefully take out the fish and place in a deep, heatproof glass dish.

Beat the eggs and squeeze the two lemons into the egg mixture.

Put this on a low heat, and slowly add the fish stock and sugar to taste.

When the mixture has thickened, pour it over the cooled fish.

Slice a cucumber thinly and use to decorate the dish.

This vintage recipe is like a good Gucci bag – it will never go out of date.

thai tuna

serves 4

4 large, fresh tuna steaks
juice of 2 limes
1 bunch of coriander,
 finely chopped
1 bunch of flat-leaf parsley,
 finely chopped
1 teaspoon lemon-grass purée
 (available in most
 good supermarkets)
½ teaspoon garlic purée

half a green chilli, chopped
 (if you like it hot, just use
 a whole chilli)
half a red pepper, finely chopped
6 cherry tomatoes,
 cut into quarters
rock salt and black pepper
 to season
olive oil

Place the tuna in a large bowl.

Squeeze on the lime juice, then add the rest of the ingredients and season to taste with salt and black pepper.

Leave to marinate for one hour.

Coat a griddle or frying pan with olive oil, heat, then add the tuna and sear each side.

Great for a girly lunch – especially if your friends are doing no carbs!

weights and measures

A Jewish Princess loves looking good. When you are thin, slim – whatever you want to call it – you're able to choose any jeans you want. If you're not lucky enough to be born with the skinny gene, then as every Princess knows, you've just got to work at it.

You can always try a tapeworm (I've heard of a supplier), or knock back a concoction of dubious herbal pills, hard-core drugs, colonic irrigation, and, as a last resort, liposuction. I've even heard of people trying to *think* themselves thin, but whenever I think of that four-letter word 'diet', I just think 'food'.

However, just like when choosing a handbag, when it comes to eating, I am *selective*. In restaurants I have been known to throw my fork down in disgust at some overly sweet or gelatinous dessert, but for fabulous food I am prepared to throw calorie caution to the wind.

Some Princesses may be in denial, but we all know that the only way to fit into those designer dresses is EXERCISE. Whether running round the shops, hitting the gym, or trying to swim with your head two feet above water (to protect your hair, of course), MOVING (yourself, that is, not changing your address) is the key to keeping your body from going south.

However, I love my food, and I am prepared to exercise my body so that I can exercise my mouth. From the number of other JPs I bump into at the gym, they've discovered this secret, too. Truth be known, I actually *love* exercise, from spinning to stepping, dancing to Pilates, you will find me doing my thaang! I find that all forms of exercise – and I mean *all* forms (except housework) – provide wonderful therapy. Exercise clears the mind, clears the cellulite and clears away the guilt that rears its ugly head whenever my fork finds that cheesecake.

All this talk of exercise has made me build up quite an appetite, so what better subject to talk about now than delicious desserts? Before we start, though, I think we should finish with one last exercise: flex your fingers and turn over the page...

7

desserts

desserts to die-t for!

I think everyone has a dessert they adore, a dessert they cannot turn down. If my son, for example, just so much as sniffs my Choca-challah Pudding (*see* page 134) baking in the oven, I'll find him prowling around the kitchen, waiting. *Challah* is delicious doughy bread made from eggs and is traditionally eaten on a Friday night. However, if there's any left over, my son gives me that puppy-dog look and I know that I will once again be whisking up eggs, chocolate and cream to make his favourite.

My own weakness is for ice cream. I just love it! I even get pleasure merely from *thinking* about it, and I get quite excited about a new special flavour. Ice cream has this unique smooth, cold, soft, melting texture that is amazing – and because it is so naughty, it is *soooooo* nice.

My relationship with ice cream once landed me in a spot of trouble.

I was entertaining some rather important clients and, due to nerves, had polished off a little too much of the pink stuff. When it came to dessert, I tottered in (high, high stilettos, naturally) with my magnificent tower of different-flavoured ice cream served in a giant brandy basket. One new, exciting flavour I had conjured up was Honey Halva Ice Cream (*see* page 138). *Halva* is a Middle Eastern delicacy made from sesame seeds which has a wonderful ability to stick to the roof of your mouth. Its unusual flavour combined with honey and made into ice cream is a winner. So it's no wonder that, in my eagerness to dive into my own dessert, I didn't wait a second before trying to serve it – and consequently could not prise the balls apart. Up shot a golden ball into the air and landed in the client's lap. I rushed over to help and found myself fumbling in said client's crotch. Scooping up the ball, I held it aloft like a triumphant egg-and-spoon-race winner and served it to his wife, who ate it. Oh dear!

Life is short. If you eat too much ice cream, life will be shorter. However, if you strike the right balance of 'moderation in everything', as my grandpa used to say, then indulging in your dessert fantasies should be encouraged, especially if it's a dessert to DIE-T for!

baked alaska

serves 6

1 box trifle sponges
marsala wine or brandy
the whites of 4 large eggs
225g caster sugar

1 x 450g carton vanilla
 ice cream
25g flaked almonds

Preheat the oven to 200°C/400°F/gas mark 6.

Place the sponges in an ovenproof dish and sprinkle with the Marsala or brandy.

Whisk the egg whites until stiff and add the sugar, whisking until the mixture is glossy.

Take the vanilla ice cream out of the freezer and leave it to soften slightly.

Smooth the ice cream over the sponges.

Spread the meringue mixture over the ice cream until it is completely covered.

If preferred, sprinkle the whole thing with flaked almonds and immediately place in the preheated oven for 20 minutes, or until the meringue mixture is lightly brown.

Serve immediately.

A great combination of cold and hot, this dessert will transport you to those snowy peaks.

black forest chocolate roulade

serves 6

1 x 435g tin black pitted
  cherries in syrup, drained
kirsh or cherry brandy
175g plain chocolate (use the
  type that contains 70% cocoa)

5 large eggs, separated
175g caster sugar
450ml double cream or non-dairy
  cream, whisked
icing sugar

The night before making the roulade, place the drained cherries in a bowl covered with your chosen liqueur and refrigerate.

Preheat the oven to 180°C/350°F/gas mark 4.

Melt the chocolate by putting it in a glass bowl over a saucepan filled with hot water.

Put the egg yolks and sugar in a mixing bowl and beat until pale. Add the melted chocolate. Beat until the mixture is smooth. Whisk the egg whites and fold into the mixture.

Pour into a Swiss roll tin lined with baking parchment that overlaps the side. Bake in the preheated oven for 15 to 20 minutes. Take out of the oven and leave to cool. Turn out onto greaseproof paper that has been dusted with icing sugar.

Whisk the cream and spread over the roulade. Drain the cherries, reserving the liqueur. With a hand blender, purée the cherries, then spread the purée over the cream.

Roll up the roulade; use the greaseproof paper, which is underneath, to help you. Place on a serving dish and decorate with icing sugar. Pour the reserved cherry liqueur in a jug to use when serving.

The seventies' dish reinvented. Very retro!

bread and butter pudding

serves 6

soft unsalted butter
orange marmalade
16 slices thick white bread,
   crusts removed
25g sultanas
3 large eggs

300ml milk
300ml whipping cream
110g caster sugar
grated rind of 1 orange
1 tablespoon demerara sugar

Preheat the oven to 180°C/350°F/gas mark 4.

Grease a 25cm ovenproof dish.

Spread the butter and marmalade on the bread and make a sandwich. Then butter the top of each sandwich.

After you have done all your sandwiches, cut them in two diagonally to create triangles.

Layer the triangles in the dish. Sprinkle on the sultanas.

In a separate bowl, whisk together the eggs, milk, cream and sugar. Pour over the triangles.

Grate the rind of the orange over the top and sprinkle on the demerara sugar.

Put the ovenproof dish inside a large roasting dish half-filled with water and place in the middle of the preheated oven. Bake for 30 minutes.

This should be served hot with cream or custard – fantastic as a real carb blow-out!

caramel ice cream

serves 6

8 tablespoons granulated sugar
4 large egg yolks

600ml non-dairy cream

In a saucepan, combine the sugar with four tablespoons of water and stir over a moderate heat until the sugar has dissolved completely. Let the sugar boil, swirling the pan occasionally until the sugar turns a light-brown colour.

Pour eight tablespoons of water into the caramel and simmer, stirring until it has melted.

Beat the egg yolks. Pour the syrup into them and continue beating until the mixture thickens.

Whisk the cream and fold into the mixture.

Pour the ice cream into a freezer-proof container and place in the freezer to set.

Try freezing the ice cream in a kugelhopf tin. When you turn it out, you'll have a hole in the middle, which you can then fill with various fruits.

cheater's cheese blintzes

serves 12

for the pancakes
(to cheat even more,
    just some buy ready-made!)
250g plain flour
2 large eggs
570ml full-fat milk
grated rind of 1 lemon
4 tablespoons oil, for frying

50g unsalted butter, to brush
    over the blintzes

for the filling
500g cream cheese
    (I use light)
6 tablespoons sour cream
2 large egg yolks
3 tablespoons caster sugar
1 teaspoon vanilla essence
50g sultanas
pinch of salt

to serve
sour cream, cherry jam or
    apple sauce

Mix all the pancake ingredients, except the oil, in a blender and leave in fridge for at least half an hour, or even better, overnight. When the above mixture is ready, make the filling simply by mixing together all the filling ingredients.

Whisk the batter before using to ensure a good consistency; you can easily do this with a hand whisk.

Coat a 20cm round, thin frying pan by pouring the oil onto it and swirling it around the pan, then pouring out any excess and wiping around the pan with a paper towel.

Heat the coated frying pan and pour on a thin layer of batter.

The pancake is ready when the mixture starts to bubble or comes away from the sides, so when that happens, just flip it over to lightly brown it. The first pancake is always a disaster, so don't worry – just eat it!

Turn the pancakes out onto a baking sheet. Continue this method, but keep oiling the pan after every two pancakes. Just stack the pancakes one on top of the other and leave to cool.

You should make approximately a dozen.

the blintz
Preheat the oven to 190°C/375°F/gas mark 5.

Take one pancake and fill it with the mixture, roll up the pancake, fold its sides and place in an ovenproof dish. Continue to do so with the others until the dish is filled.

Melt 50g unsalted butter and brush over the blintzes.

Bake in the preheated oven for 20 minutes, or until lightly browned.

Blintzes are delicious served warm with sour cream, cherry jam or apple sauce – or all three if you desire! If you're in a rush, just buy ready-made pancakes; they work equally well.

choca-challah pudding

serves 6

unsalted butter, for greasing
8 slices of a large *challah*
   1.5cm thick (one day old)
200g dark chocolate
   (the 70% cocoa type)
200ml double cream
300ml skimmed milk

115g unsalted butter
150g caster sugar
4 medium eggs

for decoration
icing sugar

Butter a medium ovenproof dish (I always use an oval one that's 35cm x 24cm x 6cm).

Remove the crusts from the *challah*. If the slices are very large, cut them in half.

Place the chocolate, cream, milk, butter and sugar into a double saucepan (bain-marie) over a low heat, stirring all the time, until the mixture is melted and smooth. If you haven't got a double saucepan, then just use a saucepan of boiling water with a heatproof bowl over it and put the ingredients in the bowl.

Remove the bowl from the heat and leave it to cool.

Beat the eggs and stir them slowly into the chocolate mixture.

Pour half the sauce into the bottom of the ovenproof dish.

Place the sliced *challah* into the liquid, pressing down with the back of a tablespoon to allow the bread to begin to saturate.

Pour over the remaining liquid and do this again to allow the sauce to saturate the bread and completely cover the *challah*.

Cover the dish and leave to cool, then refrigerate it for a minimum of two hours.

Preheat the oven to 180°C/350°F/gas mark 4.

Bake in a bain marie (I do this by putting the dish on a roasting tray and filling it with water until it reaches halfway up the outside of the dish) in the preheated oven for about 20 minutes.

Dust with icing sugar to decorate and serve warm.

A fantastic way of using leftover challah – and a wonderfully chocolicious dessert.

coffee and amaretti ice

serves 8

3 large eggs
55g caster sugar
300ml non-dairy cream

3 tablespoons coffee essence
200g amaretti biscuits

Whisk the eggs and sugar together until pale and creamy in colour, with a texture like whipped cream.

Add the non-dairy cream to the mixture and continue to whisk until thick.

Whisk in the coffee essence.

Bash the biscuits with a rolling pin to form broken pieces and fold these into the mixture.

Place the mixture into a container and leave overnight to freeze.

A gondola moment.

fruit brûlée

serves 6

seasonal fruits such as
   strawberries, raspberries
   or fresh peaches
600ml double cream

1 vanilla pod, split
4 large egg yolks
1 level tablespoon caster sugar
extra sugar, for the caramel

Make a fruit salad in advance and put it into an ovenproof dish.

To make the brûlée, put the cream and the vanilla pod into a double saucepan (bain-marie) over a moderate heat. (If you haven't got a double saucepan, use a saucepan of boiling water with a heatproof bowl over it and put the ingredients in the bowl.) Stir continuously and bring to scalding point.

Beat the four egg yolks thoroughly with the sugar.

Remove the vanilla pod from the cream and pour in the egg-yolk mixture. Turn down the heat and stir continuously for about five minutes, until the cream thickens. Keep a very beady eye on this tricky procedure; if the mixture boils, it will be ruined!

Pour this warm, thick mixture over the fruit salad. Cool. Place in the refrigerator for at least six hours (or better still, overnight) to set.

The next day, cover the brûlée evenly with about half a centimetre of caster sugar and place under a preheated very, very hot grill, turning the dish to allow the mixture to melt in order to caramelize. Make sure the top is evenly browned, then serve.

For this procedure, you can use a chef's blowtorch – but watch out for your nails!

honey halva ice cream

serves 8

3 large eggs
50g caster sugar
300ml non-dairy cream

225g *halva* – plain or any other
  flavour of your choice, crumbled
3 dessertspoons clear honey

Whisk the eggs and sugar until pale cream in colour, with a texture like whipped cream.

Whisk the cream until thick and then fold the crumbled *halva* into the cream.

Fold the egg mixture gently into the cream mixture, then drizzle in the honey and fold it in gently.

Put the mixture into an ice-cream bomb container (or any container suitable for freezing) and leave overnight in the freezer.

For those who have a sweet tooth, grab a spoon and indulge!

A Jewish Princess knows that
being happy keeps you young,
so I suggest a fabulous anti-ageing
cream. You can buy it everywhere,
it comes in huge pots, it's very
economical and it's always
available when you are feeling
a little down.

The name? **Ice cream.**
Just don't apply too much!

meringue malibu roulade

serves 6

5 large free-range egg whites
225g caster sugar
110g desiccated coconut
icing sugar, for dusting

225ml whisked non-dairy cream
  (or the real mccoy)
1 tablespoon malibu
200g chopped pineapple

Preheat the oven to 160°C/325°F/gas mark 3.

Whisk the egg whites until stiff, then add the caster sugar. Whisk until the mixture forms thick peaks.

Fold in the desiccated coconut.

Line a roulade tin with baking paper.

Spoon the mixture onto the baking paper and spread it evenly with a spatula. Bake in the preheated oven for 30 minutes.

Cover with a damp teatowel and leave to cool for at least two hours; the teatowel keeps the cake from drying out.

Place a large sheet of baking paper on a kitchen surface and dust it with the icing sugar.

Whisk the cream with the Malibu.

Turn the roulade onto the baking paper and spread it with the Malibu cream.

Place the pineapple chunks all over the cream.

Take the ends of baking paper and start to roll. DO THIS ALONE
SO YOU DON'T FEEL PUT UNDER PRESSURE!

Place the roulade into an oblong dish and decorate with any
leftover pineapple.

*If cracks appear, don't worry – that's how it's supposed to look!
This is a surprisingly sophisticated desert, but easy to do and
well worth the effort.*

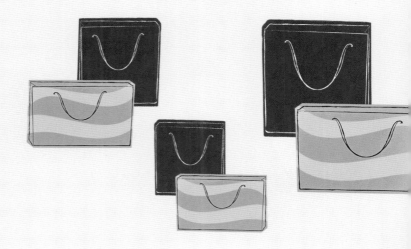

pecan pie

serves 6

225g shortcrust pastry
3 large eggs
175ml golden syrup
1 tablespoon grand marnier

pinch of salt
110g soft brown sugar
200g pecans, shelled

Preheat the oven to 180°C/350°F/gas mark 4.

Line a 22-24cm flan dish with the pastry.

Fill with baking beans and bake in the preheated oven for ten minutes.

Leave to cool and remove the beans.

Cream together the eggs, golden syrup and Grand Marnier.

Add the pinch of salt and brown sugar and mix well.

Fold in the pecans.

Pour the mixture into the pastry case and bake for 20 minutes, or until the filling has risen and turned light brown.

Serve warm or cold.

For extra calories, add a dollop of thick cream or vanilla ice cream.

shikerer's tipple

serves 6

2 litres vanilla ice cream
500g dried pitted
   prunes, chopped

110g sultanas
225ml brandy
75g plain chocolate drops

Soak the prunes and sultanas in the brandy for an hour.

Take the ice cream out of the freezer and allow it to thaw slightly.

Place the ice cream, prunes, sultanas and liqueur into a mixer and blend well.

Stir in the chocolate drops.

Pour the ice-cream mixture into a freezer-proof serving dish, and place in freezer to set.

This is a great accompaniment to any pudding or pie.

sticky briti pudding

serves 6-8

2 x 415g tins pears, in juice
(reserve the juice of 1 tin)
65g unsalted butter
200g dark-brown sugar
2 large eggs
250g self-raising flour
125ml dark rum
1 teaspoon bicarbonate of soda

for the sauce
300ml double cream
200g dark-brown sugar
75g unsalted butter

Preheat the oven to 180°C/350°F/gas mark 4.

Butter a 20cm ovenproof dish. Put one tin of drained sliced pears at the bottom of dish.

In a large bowl, mix together the butter, sugar, eggs and flour.

Put the other tin of drained pears, the rum and 125ml pear juice in a saucepan and simmer for 4 minutes.

Remove from the heat and allow to cool.

Stir in the bicarbonate of soda.

Add the pear mixture to the butter mixture.

Pour into the ovenproof dish and bake for 30-40 minutes in the preheated oven.

Check that the pudding is cooked through by putting a knife in the middle; if it comes out clean, it's done.

To make the sauce, mix all the sauce ingredients in the saucepan and cook gently until the sugar has dissolved.

Make a few holes in the pudding with a knife or other sharp instrument.

Pour the sauce onto the pudding, allowing it to seep through, and serve.

You can double up the mixture and freeze one for a later date. Alternatively, put the sponge in the freezer and defrost when ready to use. You can then make the sauce while reheating the sponge, which will take 20 minutes in a hot oven. Then just pour on the sauce as described above.

summer fruit pavlova

serves 8

the whites of 4 large eggs
225g caster sugar
225ml double cream
   (or non-dairy version)

4 kiwi fruits, for decorating
   (or any other fruits in season)

Preheat the oven to 140°C/275°F/gas mark 1.

Make a template by drawing a soft pencil line around a large round plate on a baking sheet.

Whisk the egg whites until stiff and add the caster sugar slowly while still whisking. Carry on whisking until the mixture resembles stiff peaks.

Use a spatula and shape the mixture onto a baking sheet to form a round pavlova.

Bake in the preheated oven for about one hour 30 minutes, then switch off the heat and leave the pavlova to cool inside the oven – the longer, the better.

When ready to serve, whisk the cream and spread it over the pavlova, and decorate with the fruits of your choice.

You can change the fruits according to the season – a bit like your clothes.

ten-minute trifle

serves 8

2 shop-bought raspberry
  swiss rolls
1 small bowl filled with sherry
2 x 450g tins
  raspberries, drained
2 large handfuls of roasted
  flaked almonds

140g fresh sweet raspberries
450ml fresh thick custard
300ml double cream
all of the above ingredients are
  approximate – it really depends
  on the size of the bowl you are
  going to use

Slice the Swiss rolls and dip the pieces into the bowl of sherry until the cake is saturated.

Line the bottom and sides of a glass dish with the Swiss roll.

Pour on the drained raspberries and a handful of almonds.

Next, add the fresh raspberries.

Pour the custard over the fruit.

Place in the fridge.

Whisk the cream.

Dry pan-fry the rest of the almonds until golden.

Spread the cream on top of trifle and sprinkle with the toasted almonds.

This may take only ten minutes to make, but your guests will be impressed.

tiramisu

serves 6

4 large eggs, separated
175g caster sugar
500g mascapone cheese
2 boxes boudoir biscuits
  (approximately 44 biscuits)

approximately 300ml marsala
approximately 2 tablespoons
  coffee essence
cocoa powder

Beat together the egg yolks and sugar, then add the cheese and beat until it forms a smooth, creamy mixture.

Whisk the egg whites until soft peaks form.

Slowly whisk the egg whites into the cheese mixture.

In a shallow bowl, pour in the wine and coffee essence.

Quickly dip a boudoir biscuit one at a time into the wine mixture and place at the bottom of a glass dish about 30cm x 20cm.

Place the biscuits next to each other to form a line of a single layer.

Pour a layer of the cheese mixture over these biscuits, then place a layer of quickly dipped biscuits over the cheese mixture, and finally pour on the remaining cheese mixture.

Sieve the cocoa powder over the top of the dessert and put it in the fridge to set.

The Jewish version of the Italian mama's cheesecake.

A waiter came to take the
table's dessert order.
The Jewish Princess fancied a
little bit of everything – so when
her turn came, she knew exactly
what to request:

'A fork, please!'

toffee apple pie

serves 8

for the base
75g melted unsalted butter
225g digestive biscuits, crushed
   (put the biscuits in a bag and
   bash them with a rolling pin)

for the toffee
400ml condensed milk
50g caster sugar
50g unsalted butter

for the filling
350g eating apples,
   peeled and quartered
1 tablespoon maple syrup

for the topping
300ml double cream,
   lightly whipped
1 crunchie bar

Mix the melted butter with the crushed biscuits.

Press into the base of a 21cm springform tin and chill for ten minutes.

Put the condensed milk, sugar and butter in a non-stick saucepan. Place the saucepan over a low heat, stirring constantly.

Once the butter has melted, increase the heat and bring the mixture up to boiling point, stirring constantly, and lifting the saucepan off the heat to stop the sauce from burning. The mixture will thicken; this should take no longer than five minutes.

Pour the mixture over the base and return it to the fridge.

Place the apple pieces in a microwave-safe dish.

Cook on a high heat in the microwave for six minutes.

Halfway through, remove the dish from the microwave, stir the apples and then replace to ensure even cooking.

Drain and mix with the maple syrup.

Place the syrupy apples on top of the toffee and put the pie back in the fridge.

Whisk the cream and spread it on top of the pie, then decorate with broken pieces of the Crunchie bar.

To ensure the toffee apple pie stays in one piece, use a sharp knife and run it around the edge of the tin before removing it.

8

cakes

cakes just like grandma made

My grandmother made unbelievable cakes. Her cakes were actually legendary, and to this day my friends still reminisce about the after-school tea parties we had.

Whenever she used to visit, she always arrived with pallets and pallets of farm-laid eggs, crates and crates of fruit and other bulk purchases from the cash-and-carry. I think this bulk-buying habit was due to the war years and her fear of not having enough food. My mother and I seem to have followed suit and are always overstocking our fridges and freezers ('just in case').

Grandma would spend her visits glued to the food mixer in the kitchen, baking and baking – and of course I liked nothing better than to spend time with her, helping, licking the bowl, watching and sniffing the delicious aroma of *cake*.

I have to confess that I was a social saddo in my early teens; yes, even Jewish Princesses can have some 'backward' years! My social life was practically nil. While my friends had boyfriends and were partying their nights away, my family had moved – and we were, as far as I was concerned, too far from civilization. My Saturday nights were spent, I'm almost ashamed to say, *baking*.

I realize now, however, just how incredibly therapeutic baking is. I am sure my grandma used to bake so much to get over the loss of my beloved grandpa. I, in turn, baked so much to cope with yearning for my knight in shining armour to turn up on his white charger. Amazingly, he eventually did turn up: not 'on', but 'in' a white Honda Prelude – nearly Princess Perfect!

So forget yoga or Pilates; baking is an incredible way to escape your troubles and get lost in a world of flour, eggs and sugar. To see such basic ingredients transformed into treats as wonderful as the cakes in this chapter is always a truly satisfying experience.

almond cake

225g softened unsalted butter
175g caster sugar
4 medium eggs, separated
1 pinch of salt
175g self-raising flour
50g ground almonds
1 teaspoon baking powder
1 teaspoon almond essence

for the icing
approximately 2 tablespoons
 boiling water
225g icing sugar
25g flaked almonds

Preheat the oven to 160°C/325°F/gas mark 3.

Cream the butter and sugar until pale.

In another bowl, whisk the egg whites until they form soft peaks (add a pinch of salt).

Mix the self-raising flour with the ground almonds and baking powder.

Add the dry ingredients to the creamed butter and sugar.

Slowly add the egg yolks and almond essence.

Fold in the egg whites.

Line or grease a loose-bottomed 21cm tin.

Bake in the preheated oven for 35 to 40 minutes.

Cool before turning out onto a wire rack.

To make the icing, slowly stir the boiling water into the icing sugar.

Place the icing in a double bowl (bain-marie) and, on a low heat, keep stirring until the sugar has melted and the icing is easy to spread. (If you don't have a bain-marie, place the ingredients in a bowl and put this over a pan of boiling water.)

Heat a frying pan and dry-fry the flaked almonds.

Once you've iced the cake, decorate it with the toasted almonds.

Don't feel too guilty when eating this treat, as almonds are great for the skin!

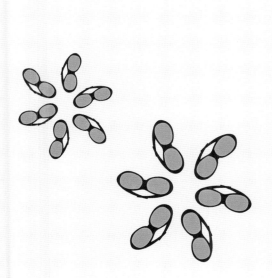

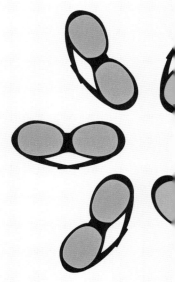

anytime cake

225g unsalted butter
225g caster sugar
4 medium eggs
1 teaspoon vanilla essence
225g self-raising flour

optional extras
110g sultanas
or 110g chocolate chips
or 110g glacé cherries...
   or all three!

Preheat the oven to 180°C/350°F/gas mark 4.

Cream the butter and sugar, then add the rest of the ingredients – and then add an optional extra to make it into the cake you fancy to break your diet on!

Pour the batter into a greased 21cm baking tin.

Bake in the preheated oven for 45 minutes to an hour, testing with a knife to see if it has cooked through.

The cake mixture can also be divided into muffin cases. Bake them for ten minutes, then decorate with icing sugar and sweets for fairy cakes.

apple brûlée cake

450g eating apples,
  peeled and sliced
110g self-raising flour
1 teaspoon baking powder
110g caster sugar
6 tablespoons milk
4 tablespoons unsalted
  butter, melted
2 medium eggs
1 teaspoon cinnamon

for the topping
75g unsalted butter, softened
110g caster sugar
1 teaspoon vanilla essence
1 egg

Preheat the oven to 160°C/325°F/gas mark 3.

Put the thinly sliced apples into the base of a greased and floured 21cm loose-bottomed tin.

Put the rest of the cake ingredients into the bowl and beat until smooth.

Pour the batter over the apples, spreading it evenly, and bake in the preheated oven for 30 to 40 minutes, or until lightly golden.

Meanwhile, cream the topping ingredients.

Remove the cake from the oven and spoon over the topping.

Bake for a further 20 to 25 minutes, until the topping is a golden brown.

This cake is delicious served hot or cold. The way to my husband's heart is with this cake – but he's taken!

as sweet as channie 'honey' cake

450g self-raising flour
225g caster sugar
grated rind and juice of
  1 large orange
1 teaspoon mixed spice
1 teaspoon cinnamon

½ teaspoon bicarbonate of soda
225ml corn oil
3 medium eggs
450ml golden syrup
225ml boiled water
flaked almonds, for decorating

Preheat the oven to 180°C/350°F/gas mark 4.

Sieve the flour.

In a bowl, mix all dry ingredients together – *i.e.* flour, sugar, orange rind, mixed spice, cinnamon, bicarbonate of soda.

Combine all the wet ingredients in another bowl then add them to the dry ingredients. Mix together.

Pour the mixture into a 25cm square cake tin and decorate with flaked almonds if desired.

Bake in the preheated oven for 50 to 60 minutes, always checking towards the end.

Even though there is no honey in this cake, it is still considered 'honey' cake by JPs. Honey cake is always made for the Jewish New Year to wish friends and family sweet success for the coming year. This one is delicious served warm, sliced and buttered.

banana cake

225g softened unsalted butter
250g caster sugar
2 large egg yolks
225g plain flour
1 teaspoon baking powder

½ teaspoon salt
4 tablespoons sour cream
2 medium-sized ripe bananas
1 teaspoon vanilla essence
5 large egg whites, whisked

Preheat the oven to 180°C/350°F/gas mark 4.

Mix the butter and sugar together.

Add the egg yolks to the mixture and continue to beat.

Add the flour, baking powder, salt, sour cream, bananas and vanilla and blend well.

Fold in the whisked egg whites.

Pour the batter into a greased 20cm square tin and bake in the preheated oven for 45 minutes.

This is great served warm with butter. Even if people don't like bananas, they will be surprised at how delicious this recipe is.

a word about cheesecake

There are many reasons why I love making cheesecake.

1   It's a fantastic stress release when you bash the hell out of those biscuits (are your children getting on your nerves?), plus it's quite a good arm workout.

2   It usually qualifies for the fewer-than-ten-ingredients rule.

3   It always comes out of the oven completely level, so it looks highly professional.

4   It takes ten minutes to prepare and 30 minutes to cook so I can still go to my Sunday spinning class.

5   What really encouraged me in my baking of cheesecake was when I gave a cheesecake to one of my husband's clients (very, very famous artists) and they said (just wait for this – it's very exciting and great for my Princess ego) that it was: 'The BEST cheesecake they had ever tasted.'

Do you think if I made them enough cheesecake they would give me a 'piece' of their work?

Every Jewish Princess knows
that cakes have no calories if you:
* take a bite from somebody
  else's plate;
* stand on one leg when eating;
* sniff them;
* only eat the top;
* only eat the bottom;
* only ever nibble a sliver!

If you're suffering with PMT
(Princess Menstrual Tension),
of course, then anything goes!

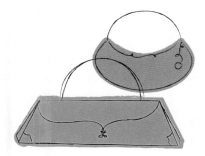

peanut butter cheesecake

8 digestive biscuits
2 tablespoons smooth
    peanut butter
1 tablespoon unsalted butter
175g caster sugar
4 medium eggs
450g curd cheese
2 tablespoons sour cream

peanut butter topping
3 tablespoons smooth
    peanut butter
2 tablespoons sour cream
1 tablespoon single cream
caster sugar to taste
a handful of crushed
    salted peanuts

Preheat the oven to 150°C/300°F/gas mark 2.

Crush the biscuits by putting them into a bag and bashing them with a rolling pin.

In a small saucepan, melt the peanut butter and butter, then add the crushed biscuits and mix well.

In a greased, round springform tin, approximately 21cm in diameter, press the biscuit mixture down to form the base.

Beat together the sugar and eggs until the mixture is light and frothy.

Add the curd cheese and sour cream.

Cook for approximately 20 minutes in the preheated oven, or until the edges are firm but the middle is slightly wobbly.

Turn off the oven and let the cake cool inside with the door slightly open.

When cold, refrigerate it for at least two to three hours.

To make the topping, melt the peanut butter in a saucepan on a low heat, then add the sour cream and the single cream – DO NOT let the mixture boil!

As soon as these ingredients are stirred in, remove from the heat and add enough sugar to suit your own taste.

When the topping is cool, remove the cheesecake from the fridge and spread the peanut butter topping on top.

Garnish with a handful of crushed salted peanuts.

If you like peanut butter, this cheesecake is orgasmic!

the ultimate cheesecake

for the base
450g digestive
  biscuits, crushed
1 teaspoon cinnamon
110g melted
  unsalted butter
75g caster sugar

for the cake
75g sultanas
2 large eggs
3 tablespoons milk
550g cream cheese
75g caster sugar
1 teaspoon
  vanilla essence

for the topping
300ml sour cream
1 teaspoon
  vanilla essence
2 tablespoons
  caster sugar

for decoration
fresh fruit

Preheat the oven to 180°C/350°F/gas mark 4.

To make the base, mix together the base ingredients, making sure the biscuits are well-crushed, and press into a 23cm to 25cm springform tin. Use a fork to press the mixture down on the base and up around the sides until the tin is well-covered.

Bake in the preheated oven for ten minutes.

Cool, then sprinkle the sultanas over the bottom.

Mix together the rest of the ingredients, pour on top of the base and bake for 30 minutes.

Remove the cheesecake from the oven and allow it to cool for 15 minutes.

Meanwhile, make the topping by mixing all the topping ingredients together.

Pour over the cheesecake, return it to the oven and bake for a further ten minutes. Leave to cool.

Refrigerate overnight.

Pile fresh fruit such as strawberries and blueberries on top of this cheesecake. It gives it a great look and tastes delicious.

carrot cake

100g self-raising flour
1 teaspoon baking powder
½ teaspoon cinnamon
1 pinch salt
200g light brown sugar
15g olive oil
2 medium eggs
200g carrots, grated
150g ground almonds

for the frosting
175g light cream cheese
55g icing sugar
25g softened unsalted butter

for decoration
a handful of toasted pine nuts

Preheat the oven to 170°C/325°F/gas mark 3.

Sieve and mix together the flour, baking powder, cinnamon and salt.

Add the brown sugar, olive oil and eggs and mix well.

Mix in the carrots and ground almonds.

Put the mixture into a greased 20cm cake tin and bake in the preheated oven for 30 minutes.

Turn the cake out onto a wire rack and leave it to cool.

To make the frosting, use a hand blender to mix all ingredients together in a bowl until smooth.

When the cake is cold, spread the frosting on top.

Decorate with toasted pine nuts, if desired.

A cake full of vitamins that can be used to improve your suntan if eaten regularly!

chocolate refrigerator cake

225g plain chocolate
2 large eggs
2 level dessertspoons caster sugar
225g melted unsalted butter
225g plain digestive
  biscuits, crushed
110g salted peanuts

75g chopped walnuts
225g glacé cherries, chopped
1 teaspoon vanilla essence
a handful of walnut halves,
  for decoration

Break the chocolate into squares and place in a double-saucepan (bain-marie) or in a small bowl over a pan of hot water to melt.

Beat the eggs and sugar together, then beat in the melted butter.

Stir in the chocolate and fold in the broken biscuits, together with the nuts, glacé cherries and vanilla essence.

Pour the mixture into a buttered Pyrex dish 15cm to 18cm deep and leave in the fridge to set for at least six hours.

When ready to turn out, use a sharp knife and run it around the edges of the cake. Immerse the dish halfway in hot water to allow the cake to melt a bit and then turn it out.

Decorate with the walnuts and refrigerate until ready to use.

When serving, cut into small slices as this cake is very rich. Delicious used as a petite four with coffee.

de luxe chocolate cake

175g dark chocolate
(the 70% cocoa type)
110g softened unsalted butter
1 tablespoon grand marnier
5 medium eggs, separated
110g ground almonds
1 pinch of salt

75g caster sugar
2 tablespoons double cream

for the raspberry glaze
4 tablespoons raspberry jam
(with seeds)
1 tablespoon caster sugar

Preheat the oven to 170°C/325°F/gas mark 3.

Melt the chocolate and butter in a double saucepan (bain-marie), stirring constantly so that the mixture does not burn. If you haven't got a double saucepan, then just use a saucepan of boiling water with a heatproof bowl over it and put the ingredients in the bowl.

Let the chocolate mixture cool. You can speed this up by standing the bowl in cold water.

Stir in the Grand Marnier, egg yolks and ground almonds.

In a separate bowl, whisk the egg whites with a pinch of salt until soft peaks form, then slowly add the sugar.

Gently fold the egg whites into the chocolate mixture.

Add the cream and gently fold it into the mixture.

Pour into a greased and floured 20cm springform tin.

Place in the centre of the preheated oven for 25 minutes, or until a knife stuck in the middle comes out clean.

When the cake has cooled, release it and place it on a serving dish.

When the you are ready to glaze, place the raspberry jam and sugar in a saucepan and heat gently, stirring so that the glaze does not burn. When the mixture becomes syrupy, remove it from the heat.

Pour the glaze over the top and let it run down the sides.

Even when this cake is served as a dessert after a large meal, there is never any left over! If you don't like raspberry jam, simply try making the glaze with the jam of your choice.

english fruit cake with a cup of tea

250g mixed chopped
   dried fruit
grated rind of
   1 whole orange
4 english breakfast tea bags
600ml hot water
3 large eggs
110g caster sugar
110g softened unsalted butter
2 tablespoons milk

280g self-raising flour,
   plus 3 tablespoons
1 teaspoon baking powder

for decoration
1 handful of blanched almonds
1 tablespoon demerara sugar

Preheat the oven to 150°C/300°F/gas mark 2.

In a large heatproof bowl, place the dried fruit and orange rind.

Make a pot of tea with the four tea bags and boiling water and leave it to brew.

Pour the tea over the fruit until it is just covered and leave to marinate for a couple of hours.

Whisk the eggs and sugar together until the mixture has expanded and is nice and frothy.

Add the butter slowly, in small pieces.

Strain the fruit, reserving the tea, and dry it on kitchen paper, then lightly dust it with the three tablespoons of flour.

With the mixer set on slow, add the fruit, milk and the reserved tea to the egg, sugar and butter mixture.

Sift together the self-raising flour and the baking powder, then add this to the cake mixture.

Grease and flour a 20cm springform tin and pour in the batter.

Decorate with a handful of blanched almonds and sugar.

Bake in the preheated oven for 55 to 60 minutes.

A very useful aprés-school-run cake. This is best made the day before serving.

grandpa's favourite coffee and walnut cake

225g unsalted butter
225g caster sugar
4 large eggs
110g chopped walnuts
2 tablespoons coffee essence
225g self-raising flour
2 teaspoons baking powder

for the icing
175g unsalted butter
450g sifted icing sugar
4 tablespoons milk
6 teaspoons coffee essence

for decoration
walnut halves and coffee beans

Preheat the oven to 160°C/325°F/gas mark 3.

Mix all the cake ingredients together and divide into two greased 18cm baking tins.

Bake in the preheated oven for 25 to 30 minutes, checking that the sponges are baked by putting a skewer in. If it comes out clean, they're done.

Remove the cakes from the oven and leave them to cool.

Make the icing by combining all the ingredients, then use some of it to sandwich the sponges together.

Ice the cake all over, and decorate with walnut pieces interspersed with coffee beans.

The best present for Grandpa on birthdays, anniversaries or any occasion.

honey cake

225g plain flour
115g caster sugar
1 teaspoon cinnamon
1 teaspoon mixed spice
50g clear honey
115ml golden syrup

50ml cooking oil
2 medium eggs
1 teaspoon bicarbonate
  of soda
80ml smooth orange juice

Preheat the oven to 170°C/325°F/gas mark 3.

Grease and flour a 20cm cake tin.

In a large bowl, mix together the flour, sugar and spices.

Add the honey, syrup, oil and eggs.

Beat well until smooth.

Use another large bowl and dissolve the bicarbonate of soda in the orange juice. Stir well (this will fizz) and then add this to the mixture.

Bake in the preheated oven for 40 to 60 minutes.

It is a good idea to make this cake the day before eating. As it matures, it becomes even more syrupy and delicious.

marble cake

3 large eggs, separated
175g unsalted butter
200g caster sugar
1 teaspoon vanilla essence
225g self-raising flour
3 tablespoons milk

75g plain chocolate drops
    (or milk chocolate, if you prefer)
4 tablespoons plain chocolate
    spread (or milk chocolate,
    if you prefer)

Preheat the oven to 180°C/350°F/gas mark 4.

Whisk the egg whites until stiff. Leave them in the bowl and put to one side for later.

In a separate bowl, beat the egg yolks, vanilla essence and butter until creamy. Add the sugar gradually, beating until the mixture resembles whipped cream.

Now comes the tricky part. Keep the mixer going on low and add to it a tablespoon of flour and alternate with a tablespoon of milk and a tablespoon of the egg whites that you whisked at the beginning. Keep doing this until all the ingredients have been used. (This is a bit of a pain, but really worth doing as it makes the cake very light.)

Divide the mixture into two. Place one half into a new bowl and add the chocolate drops to it.

Pour the mixture containing the chocolate drops into a greased 20cm baking tin and spread it out over the bottom.

Add the chocolate spread to the other half of the mixture and mix well. Pour this over the chocolate-drop mixture in the baking tin.

By making a figure of eight with a spatula, swirl the dark chocolate-spread mixture gently through the chocolate-drop mixture, so that you get a marbling effect. When you do this, lift the spatula upwards to allow the white mixture to come to the top.

Smooth the top of the mixture and bake for 45 to 60 minutes.

I can remember fighting with my brother as to whose turn it was to lick the bowl when we were children. Sometimes for children (and even a few adults!) this cake mixture is even more delicious than the finished cake. So leave the bowl for the kids – or if they're at school, add washing up liquid and quickly fill it with water to prevent you from licking.

marmalade cream cake

175g softened unsalted butter,
  cut into small pieces
175g caster sugar
3 medium eggs
1 teaspoon
  orange-flower water
3 tablespoons full-fat milk
1 tablespoon warm water
175g fine self-raising flour
2 teaspoons baking powder

for the marmalade-cream filling
300ml whipping cream
3 tablespoons orange
  marmalade (thin-cut)

for decoration
icing sugar
grated rind of 1 whole orange

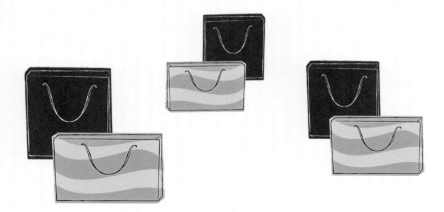

Preheat the oven to 160°C/325°F/gas mark 3.

Grease and flour two 20cm sandwich tins.

In a large bowl, cream the butter and sugar together until pale.

Add the eggs, one at a time, beating after each addition.

Add the orange-flower water, and beat it in.

Mix the milk with the warm water, then add to the mixture and continue to beat.

Add all the dry ingredients and mix thoroughly.

Pour into the prepared sandwich tins and bake in the preheated oven for approximately 20 minutes.

Once the cake has cooled, turn it out onto a wire rack.

To make the filling, whisk the whipping cream until firm.

Stir the marmalade so that it is loosened up and soft, then add it gently to the whipped cream.

Fill the centre of the cake with the marmalade cream.

To decorate, grate the orange over the top of the cake and then lightly dust with icing sugar.

This is a wonderful cake to serve for a Sunday tea.

sour cream cake

75g unsalted butter
1 teaspoon cinnamon
175g (approximately 14)
   digestive biscuits, crushed
425ml sour cream

110g caster sugar
1 medium egg
1 x 110g tin raspberries, drained
fresh raspberries, for decoration
more sour cream, for serving

Preheat the oven to 150°C/300°F/gas mark 2.

Grease a 22cm springform cake tin.

Melt the butter in a saucepan over a gentle heat. Add the cinnamon, stir, then add the cinnamon-butter to the crushed biscuits in a large bowl.

Press the mixture into the bottom of the baking tin to make a base.

Bake in the preheated oven for five minutes.

Whisk together the sour cream, sugar and egg.

Add the well-drained tinned raspberries.

Pour the sour cream mixture over the base and bake for 40 minutes, or until set.

Remove the cake from the oven. When cool, refrigerate it in the tin for at least an hour to allow it to set.

Remove the cake from tin right before serving.

Decorate with fresh raspberries and serve with a little sour cream on the side.

Be warned: one slice is never enough!

whisky cake

finely grated rind of
  1 large orange
4 tablespoons whisky
175g sultanas

175g unsalted butter
175g caster sugar
3 large eggs
225g self-raising flour

Soak the grated orange rind in the whisky for at least an hour, then add the sultanas and soak for a further 45 minutes.

Preheat the oven to 180°C/350°F/gas mark 4.

Grease and line a 20cm round cake tin with baking parchment.

In a large bowl, cream together the butter and sugar until light in colour, then add the eggs.

Fold in the flour.

Add the sultanas and whisky mixture.

Pour the batter into the prepared tin and bake in the middle of the preheated oven for one hour.

Cool completely before turning out and serving.

When the nights are drawing in, a slice of this is great with a cup of tea.

zesty lemon cake

4 medium eggs
225g caster sugar
225g softened unsalted butter
2 tablespoons milk
1½ teaspoons
  lemon essence
grated rind of
  1 whole lemon
225g self-raising flour
1 teaspoon baking powder

for the lemon syrup
juice of 2 lemons
grated rind of
  1 lemon
grated rind of
  1 orange
6 tablespoons water
16 tablespoons caster sugar

Preheat the oven to 180°C/350°F/gas mark 4.

In a large bowl, whisk the eggs with the sugar until the mixture is frothy and the sugar is fully dissolved.

Add the softened butter, slowly, in small pieces.

Add the milk and lemon essence.

In another bowl, mix the rind of one whole lemon with the flour and baking powder, then add this to the egg mixture.

Pour the mixture into a greased and floured 20cm x 20cm square springform tin.

Place the tin in the middle of the preheated oven and bake for 45 minutes.

To make the syrup, place all of the syrup ingredients in a saucepan.

Bring to boiling point and keep it there for at least three minutes, sitrring constantly, until the syrup has gone a golden colour.

As the cake comes out of the oven, prick it all over and pour the syrup over the top. It looks lovely with the mixture of lemon and orange rind.

When the cake is cool, lift it out of the tin.

This cake is a combination of sweet and sharp, chewy and spongy – and absolutely delicious!

9

the ultimate dinner

the ultimate friday night dinner

Friday night. The Sabbath or *Shabbos*. The night the family sits down together to share a wonderful meal. This is the night when the house is filled with incredible smells of cooking. The smell of chicken soup permeates every room.

The table is laid with a crisp, white tablecloth. The *Shabbos* candles sit in their treasured silver candlesticks. The gleaming silver *kiddush* cup filled with *kiddush* wine is ready for the blessing, and two wonderful freshly baked *challas* sit side by side, covered with a beautifully embroidered cloth.

This is the night when we reconnect with our spiritual being and are thankful for the week and the Sabbath still to come. For Jews, the Friday night dinner is a bit like Christmas – except instead of having to make this feast once a year, we have the blessing of making it once a week! The Friday night dinner fantasy is one of Father returning home from work in plenty of time before sundown, bearing flowers for his beautiful wife. The children are in their best clothes, faces shining like little angels, while Mother makes the blessing over the candles, looking like she has taken a break from the film set of *Fiddler on the Roof*. The house is serene, the conversation around the table jovial, interesting and happy.

Unfortunately, I have since discovered that the fantasy of the Sabbath meal is, of course, nothing like reality.

Reality is: Husband returns home from work very, very late and flowerless. Your teenager has to be prised off the phone and strapped into her seat at the table, while the younger children, stuck with masses of homework, are moody and tired from a long week of school. Mother is – well, of course being a JP, *she's* perfect! But as far as the conversation around the table goes, if you're not careful, it can wind up turning into World War III.

Over the years, I have found ways of dealing with Friday night dinners and ways of turning the reality into at least a little bit of the fantasy. Firstly, I always try to invite guests over. On the face of it, this seems insane, but like most things in my life, there is method in my madness. Even though I now have more work to do, there are many advantages. For one thing, my husband magically gets home on time (even if he shoots in just before the guests arrive) – and sometimes he's even bearing flowers.

The teenager looks respectable and manages to unglue the phone from her ear (even teenagers have a social conscience).

The younger children are excited that guests are coming over and even enjoy showing off by helping to clear up.

Guests always lighten the atmosphere, too, so there are never any arguments around the table – plus they are actually *grateful* for my culinary delights. An additional bonus is that we invariably get invited back, so there's one less Friday night dinner for me to do.

I have also learned to prepare the Friday night dinner early. Chicken soup is cooked on Wednesday, the table is laid and desserts made on Thursday, and on Friday I only allow one hour to prepare the rest of the food.

One hour may seem a short amount of time to prepare this feast, but as I have learned, you can prepare the same amount of food in one hour that used to take you days. It's just a matter of self-belief – and making things simple.

If, one week, you are too tired or too busy to do the full works, either cheat (otherwise known as 'buy in') and make it look like your own, or simply simplify. No one will think any more or any less of you, and nobody wants to be greeted by a growling, resentful Princess hostess.

I am sure you have questioned why we even do Friday night dinner, but as Topol from *Fiddler on the Roof* said, 'Tradition! Tradition!'

The Jewish Princess has a long
list of things she has to prepare
in order to be ready for the
Sabbath: manicure, pedicure,
hair colour and cut, facial
and, of course, medication –
I mean, *meditation!*

chopped liver

serves 8

1.3kg onions, roughly chopped
(you can use frozen
ready-sliced)
4 tablespoons corn oil
450g ox liver

4 large hard-boiled eggs
(reserve 1 for decoration)
3 slices brown bread,
crusts removed
salt and black pepper to taste

Liver is the only meat you buy from a kosher butcher that you need to kosher yourself (with special koshering utensils: ask a kosher butcher for advice). To do this, wash the liver to remove any blood, cover it with coarse salt, broil it on a open flame, then rinse it thoroughly.

In a large frying pan, fry the onions in the corn oil until they turn a dark brown. The browner the onions get without burning, the better the result.

Cut the liver into strips and add it to the fried onions. Cook very slowly until the liver is soft and brown all the way through. Add a little more oil if necessary.

When the onions and liver are fully cooked, add the rest of the ingredients and blend well. Add salt and pepper to taste, but be sure to mix the seasoning thoroughly into the mixture.

Place in a serving dish. Using a coarse (large-holed) grater, grate the reserved egg over the top to decorate, and refrigerate.

If this is not for you, try buying it at your local kosher butcher – but I bet it won't taste half as good as this!

egg and onion

serves 8

10 medium hard-boiled eggs
1 bunch of spring onions,
  finely sliced
olive oil

salt and black pepper to taste
1 avocado (optional)
salad leaves and chopped
  tomatoes, for serving

Using a coarse (large-holed) grater, grate the hard-boiled eggs and stir in the sliced spring onion.

Add enough olive oil to bind the mixture.

Add salt and pepper to taste.

For a different twist, add the avocado: peel it and mash it with a bit of lemon juice and mix it into the egg mixture.

Use an ice-cream scoop to make individual servings. Place each serving on a bed of lettuce leaves and chopped tomatoes.

a word about chicken soup

Of all the soups in the Jewish kosher diet, none is more well-known or well-loved than chicken soup. And with good reason: this soup is traditionally served on Friday nights as part of the *Shabbos* dinner, and it is *wonderful*.

Out of all the chicken soups I have ever tasted, however, no two taste the same. Even when a recipe is handed down from mother to daughter, it still does not taste the same.

This is the Mystery and Magic of chicken soup: every recipe has something unique to offer.

I love my own chicken soup because it is full of intense flavour, it is low-fat, and it takes very little time to prepare. In fact, I always make my soup on a Wednesday evening so that I can leave it overnight to cool; this seems to intensify the flavour. Refrigerate it on Thursday morning, then just reheat it for Friday night.

Another handy tip I've discovered concerns *lockshen*: place a heaped tablespoon of cooked *lockshen* into the bottom of each individual soup bowl. As you pour the soup in, the *lockshen* will heat through.

Once you've got used to making chicken soup (whether by following this or any other recipe), you'll probably want to experiment with using other root vegetables or herbs to make your own version of this delicious elixir.

chicken soup

serves 10

1 fowl, skinned and cut into 8 pieces
NOTE: A fowl is a female chicken that's a bit like a mother-in-law – e.g. a 'tough old bird'. It can be purchased from any kosher butcher (the fowl, not the mother-in-law!).
1 bunch of celery hearts, washed
8 carrots, peeled and left whole
4 turnips, peeled and cut into large chunks

2 parsnips, peeled and cut into large chunks
1 swede, peeled and cut into large chunks.
1 large onion, peeled and cut into large chunks.
2 tablespoons dried parsley
salt and black pepper to taste
1 tablespoon chicken-flavoured soup mix to every 600ml water used

Clean the fowl, removing any excess fat and anything else that looks vaguely suspicious.

Place it in a very large *shissel* (cooking pot).

Add all the rest of the ingredients and cover with cold water.

Bring the soup to the boil. As there is no skin on the fowl, YOU WILL NOT NEED TO STAND THERE FOR HOURS REMOVING THE SCUM – so this is a perfect recipe for Princesses.

Turn the heat down to low, cover the saucepan and leave to simmer gently for at least five hours, or more. You can test when the soup is ready by using a fork – the chicken should be so soft that it should practically fall away from the bone, and the soup should be a deep golden colour.

to serve

If you have made this in advance, then when you take the soup out of the fridge to serve, get rid of any fat that has risen to the surface by laying several sheets of kitchen paper on top of the soup. When you remove it, the fat will cling to the paper.

If you're not worrying about your carbs, it's traditional to serve this soup with matzo balls and egg noodles (*see* Matzo Balls, page 196). Egg noodles can be purchased in any kosher section of the supermarket or delicatessen, but if you cannot be bothered to *shlep* around, just add some cooked pasta of your choice.

Finally, if you are lucky enough to have the golden eggs (*see* page 81), here's how to cook them. First, wash the eggs and peel off the outer layer. When you're ready to serve your chicken soup, just pop the eggs in the soup (also the matzo balls and *chicklach; see* page 196-7) and bring it to boiling point. Let it boil for five minutes to ensure that the eggs are cooked through.

Ahhhhhhh: Jewish Princess penicillin!

matzo balls (knaidlach)

serves 10

7 large matzos
1 large onion, diced
1 tablespoon olive oil
3 large eggs

3 tablespoons fine matzo meal
1 dessertspoon
    chicken-flavoured seasoning
salt and pepper to taste

Break the matzos into small pieces and place them in a colander. Soak with water. Leave to drain.

Fry the onion in the olive oil until brown.

Place in a mixer the matzos, eggs, matzo meal, onions, chicken flavouring and salt and pepper. Mix all the above until you can see the mixture binding together.

Bring a large saucepan of salted water to the boil.

Wet your hands, then take a teaspoon of the mixture and roll it into a ball.

Continue to make the rest of the balls and then place them in the boiling water for 20 minutes.

This is a great accompaniment to Chicken Soup (see page 194) and is a bit different to the usual knaidlach.

chicklach (kosher-style dim sum)

serves 10

450g minced chicken
2 medium eggs
140g medium matzo meal
6 tablespoons chicken fat
   (use the fat that comes off the
   top of the chicken soup – *see*
   page 194) or dairy-free
   margarine, melted

½ teaspoon cinnamon
½ teaspoon ginger
1 teaspoon dried parsley
Salt and pepper to taste

Mix all the ingredients together and leave the mixture in the fridge
to firm up for 20 minutes.

Wet your hands and, using a teaspoon as a measure, form the
mixture into small balls.

Fill a large saucepan with salted water and bring it to the boil.

Drop the *chicklach* into the water and simmer for 30 minutes.

Use a slotted spoon to remove the *chicklach* from the water and
add them to your chicken soup when needed (*see* below).

*This new recipe will become a must as an accompaniment to
Chicken Soup (see page 194).*

roast chicken

serves 6

1 roasting chicken
2 onions, peeled
black pepper
dried parsley or sage
olive oil

chicken seasoning powder or
    chicken stock powder
12 mushrooms, washed and
    cut into quarters.
spring onions
soy sauce

Preheat the oven to 180°C/350°F/gas mark 4.

Clean the chicken. Just before putting it into the oven, pour boiling water through the centre of the chicken and over the top to make the skin crispy.

Place the chicken in a roasting dish. Place one of the onions inside the chicken's cavity, then season the chicken with black pepper and sprinkle the parsley over the top. Rub a little olive oil into the skin to work the seasoning and parsley into the chicken.

Pour cold water into the bottom of the roasting tin (about 700-850ml depending on the size of chicken) and add a good sprinkling of chicken seasoning or stock powder.

Add the other onion, sliced, and mushrooms, then add the chopped spring onions into the water and a splash of soy sauce onto the chicken.

Cook in the preheated over for one hour 30 minutes. Lovely!

This recipe should serve six, depending on the size of the chicken. Unless my Uncle Michael is one of the guests – then it serves one...

roast potatoes

serves *as many as you like!*

roasting potatoes (allow at least
  2 large potatoes per person)

2-3 tablespoons olive oil
sea salt

Preheat the oven to 180°C/350°F/gas mark 4.

Peel the potatoes and slice them in half.

Take a roasting dish and pour two to three tablespoons of olive oil over the bottom.

On each half of the potatoes, use a sharp knife and make slits (about five) across the potato's width. The slits should be three-quarters of the way down the potato; this allows the olive oil and seasoning to penetrate.

Place the potatoes in the dish, flat-side down.

Sprinkle them with olive oil and sea salt.

Bake in the preheated oven for one and a half hours, turning halfway through.

This elevates the humble roast potato to new heights. Make sure you make plenty for seconds.

shredded red cabbage with apples and sultanas

serves 6

1 red cabbage
1 apple, grated
110g sultanas

5 tablespoons sweetener or sugar
300ml white-wine vinegar
850ml water

Preheat the oven to 150°C/300°F/gas mark 2.

Slice the cabbage into thin shreds and rinse with water.

Mix the cabbage with the apple, sultanas and sweetener and put into an ovenproof dish approximately 30cm square.

Cover the mixture with the vinegar and water, making sure there is enough liquid to cover it completely.

Cover the dish with foil and bake in the preheated oven for approximately three hours. Keep an eye on it, and stir it every half an hour.

Do this the day you're going to have a manicure, as your hands can go a little bit purple. Better still, wear gloves when chopping.

sweet potato and carrot purée

serves 8

4 large sweet potatoes
8 large carrots
1 tablespoon vegetable
   bouillon powder

salt and black pepper to taste
1 tablespoon non-dairy margarine
1 medium egg, beaten

Peel all the vegetables and cut into chunks. Place them all in a large saucepan.

Barely cover the vegetables with water and add the vegetable bouillon powder.

Bring to the boil and simmer for 40 minutes. Drain and then purée.

Preheat the oven to 180°C/350°F/gas mark 4.

Add the seasoning, margarine and egg.

Pour the mixture into an ovenproof dish.

Bake in the preheated oven for 20 minutes.

I always use a hand blender to do the purée; it saves on time and washing up!

apple strudel pie

serves 6

for the dough
450g plain flour
175g caster sugar
225g unsalted butter
  or non-dairy margarine
4 large egg yolks (reserve the
  whites for the filling, below)
2 tablespoons milk
  (or soya milk if you want to
  keep the dessert *parev*)
1 teaspoon baking powder

for the filling
6 apples, peeled and sliced
110g icing sugar
3 teaspoons cinnamon
75g chopped walnuts
2 tablespoons
  vanilla-flavoured sugar
4 large egg whites

Make the dough from all the dough ingredients by using your hands to knead properly on a flat surface until everything is well-combined. If you don't want to get too sticky, mix the ingredients in your food mixer or processor. Cover the egg whites and place them in the fridge for use later.

Wrap 250g of the dough in clingfilm and place it in the freezer for at least an hour, until it is completely frozen and hard.

Wrap the remaining dough in clingfilm and place it in the fridge.

When ready to proceed, preheat the oven to 180°C/350°F/gas mark 4.

Whisk the egg whites until stiff.

Place the rest of the ingredients in a large bowl and stir to make sure there is an even coverage of the sugars and cinnamon.

Roll out the dough from the fridge to whatever thickness works to cover the bottom and sides of the greased ovenproof tin you're using (I always use a 36cm by 20.5cm one).

Place the filling on top and spread the egg whites over the filling. Take out the dough from the freezer and grate it over the egg whites.

Bake for 15 to 20 minutes in the preheated oven, or until the surface is golden and baked through.

If you can't get hold of vanilla sugar, it's easy to make and keep in your store cupboard. Just take two vanilla pods and use a knife to split them down the middle. Cut them into 3cm pieces and fill a small empty spice jar with caster sugar and vanilla. I would wait a week to allow the vanilla to infuse into the sugar. When you are ready to use the sugar, dispose of the vanilla pods.

This is such a delicious dessert that even though your guests may be full, I'm sure they will still indulge in a bissel of strudel pie.

jaffa chocolate mousse

serves 6

225g plain chocolate
  (the 70% cocoa kind)
8 medium eggs, separated

8 teaspoons cointreau
grated orange-flavoured dark
  chocolate, for decoration

Whisk the egg whites until they form stiff peaks.

Melt the chocolate in a double saucepan (bain-marie) on the hob over a gentle heat. If you don't have a double saucepan, use a small bowl placed over a pan of hot water.

When the chocolate has melted, remove it from the heat.

Stir in the egg yolks one at a time.

Using a metal spoon, gently fold in one tablespoon of egg whites, then fold in the rest.

Add the Cointreau and stir gently.

Refrigerate.

Before serving, decorate the mousse with grated orange-flavoured dark chocolate.

lockshen pudding (noodle pudding)

serves 6

250g (1 packet) fine egg
   vermicelli, boiled and rinsed
   with cold water
175g sultanas
2 large eggs
1 dessertspoon cinnamon
1 cooking apple, grated and
   drizzled with lemon juice

3 dessertspoons granulated sugar
1½ teaspoons almond essence
50g non-dairy margarine, melted
2 tablespoons orange cordial

for topping
2 tablespoons sugar
1 teaspoon cinnamon

Preheat the oven to 190°C/375°F/gas mark 5.

Mix all the ingredients, except the sugar and cinnamon for topping, and put the mixture into a greased ovenproof dish.

Mix the sugar together with the cinnamon and sprinkle over the pudding.

Bake in the preheated oven for 45 minutes to one hour.

The Friday night finale.

10

useful information

the jewish princess
does yiddish

YID DISH? No, it's not a main course

YID DISH? No, it's not a good-looking Jewish guy!

YIDD-ISH? No, it's not someone who is only a little bit Jewish.

YID D.I. SH? No, it's not a Jewish detective!

yiddish is a wonderful language

The Yiddish language is like a recipe: take some German and add to it your Hebrew, season with Slavic, a teaspoon of Romanian and finally, add a little English and French for that *je ne sais quoi*.

When used, you can express yourself in words that the listener doesn't have to understand to know exactly what you mean. The words have a warmth and earthiness to them. They are also incredibly powerful, as one word will convey feelings that you could not begin to sum up in paragraphs.

Remember *ch* is pronounced as if you are trying to clear your throat. *T* at the beginning of a word is silent; for example: *tsures* ('troubles') is pronounced *sur – es*.

In the following pages I have listed some of my favourite Yiddish words and expressions, many of which you will find within this book – and many of which, if you are a Jewish Princess, I am sure you already use.

Please note: because Yiddish comes originally from the Hebrew, and thus (like Greek) from no English-letter-based alphabet, English spellings vary from dictionary to dictionary, and even from person to person. Many Princesses have their own way of spelling, depending on where they were brought up. Spelling issues aside, I hope you find this useful when you are looking for just the right word to express how you feel!

yiddish/english glossary

Alter kocker An old person who looks old (obviously). Can't be found in Beverly Hills.

Ashkenazi A Jewish person from central or Eastern Europe.

Bagel Roll with a hole in the middle, eaten a lot on Sundays.

Balabatish An excellent homemaker. A true Jewish Princess!

Bar mitzvah A ceremony marking the religious coming of age of a 13-year-old boy (very expensive party).

Bashayrt Fate.

Bat mitzvah Ceremony marking the religious coming of age of a 12-year-old girl (also very expensive!).

Bissel A very small amount.

Blintz Pancake filled with cream cheese (*see* page 132).

Borscht Beetroot soup (*see* page 63).

Borscht Belt Popular holiday resort in the Catskill Mountains in upper New York State. For the European version, go to Grand Rimini in Italy.

Boychik Young boy.

Bris Circumcision for baby boys (ouch!).

Brocheh Blessing. There is a different one for every situation.

Broyges An argument with bad feeling. Happens a lot with families.

Bubbeleh Term of endearment – or a Passover pancake.

Bubbeh Grandmother (always imagine an old-looking grandma – not a surgically enhanced Silver Surfer!).

Bubbeh-myseh Made-up story. I know some people – and they are all under 12 – who are very good at these!

Bupkes Nothing.

Challah Platted loaf used mainly for the Sabbath meal, delicious (*see* page 134 for Choca-Challa Pudding).

Challish To want something, as in 'I *challish* for it.'

Chanukah The Festival of Lights (normally falls near Christmas). Eight days of presents: a JP's favourite time of the year!

Chasid Member of an orthodox religious sect.

Chayshik (pronounced *hay – shik*) Enthusiasm.

Chazzen Singer in Synagogue (like Neil Diamond in *The Jazz Singer* if you're lucky).

Chazzer Being greedy.

Cheder A school where children learn about the Jewish religion.

Chochem A clever person.

Cholent Meat stew that is cooked overnight (*see* page 96).

Chotchkeh Knick-knack.

Chrayn Horseradish sauce. Delicious hot red sauce that accompanies fish.

Chuppah Wedding canopy.

Chutzpah Cheek or nerve. A Jewish Princess has plenty of this!

Diaspora The historical dispersion of the Jews.
Doven Pray.
Drek A vulgar expression for ugly.

Farfel Tiny noodles.
Faygeleh Homosexual.
Feh! Ugh! Used when you don't like something.
Ferbissener A bitter person.
Fershtinkiner Revolting and smelly.
Flayshedik Kosher meat dishes.
Fliegel Chicken wing.
Forspeise A taster or appetizer.
Fress Eat a lot.
Frum Religious person.
Frummer Religious person.

Gatkes Long johns.
Gedempte Slowly cooked.
Gefilte fish A dish made from chopped fish; it can be fried or boiled.
Gelt Money.
Gesundheit You say it when someone sneezes and it means health.
Get A Jewish divorce.
Gavalt Shock. People usually say *Oy gavalt!* if something really bad happens.
Glatt kosher Strictly kosher.
Glitch When something goes wrong.
Golem Artificially created man.
Gonif A thief.
Gornisht Nothing.

Haggadah A book telling the Passover story.
Haimish (No, not a Scotsman...) Friendly and warm.
Halachah Religious law.

Halva Sweets made from sesame seeds (*see* page 138 for Honey Halva Ice Cream).
Hatikvah Israeli national anthem.

Kaddish Mourner's prayer.
Kapel or *Kippot* Skullcap Jewish men wear in synagogue – and if religious they wear all the time.
Kashrus Dietary laws.
Kibbutz Israeli cooperative agricultural settlement.
Kichel A biscuit.
Kiddush Blessing recited over wine or bread on the Sabbath or at a festival.
Kinder Children.
Kishkes Guts.
Klutz Clumsy person.
Knaidlach Matzo balls (*see* page 196).
Knish Baked roll filled with potato or meat.
K'nocker A big shot. Or a big diamond.
Krank Something annoying.
Kreplach Ravioli with chopped meat.
Kugel Noodle or potato pudding.
Kvell To glow with pride.
Kvetch To complain.

Langer Lockshen Tall person, very thin – spaghetti-like.
Latke Potato pancake.
Lechayim A toast to life.
Levoyah Funeral.
Loch in kop A hole in the head.
Lockshen Noodles.
Lox Smoked salmon.
Lubavitch A religious sect of orthodox Jews.
Macher An organizer.

Machetunim Relatives by marriage.

Matzo Unleavened bread.

Matzo-brie Matzo mixed with egg to make an omelette.

Mazel Good luck.

Mazel tov Congratulations.

Megillah A long story. It derives from the story of Esther at Purim.

Menorah Candelabra lit at Chanukah.

Mentsh A good person.

Meshugga Mad.

Meshuggener A mad person.

Metsiah A bargain – like getting a Gucci handbag at 70 percent off!

Mezuzah Religious scroll in an encasement attached to a door.

Mieskayt An ugly person.

Mikvah A ritual bath.

Milchik Dairy foods.

Minyan Ten men required for religious services.

Mishegass A mad idea.

Mitzvah A good idea.

Mockers Put bad luck on something.

Mogen Dovid A Star of David.

Mohel The man who performs the circumcision (ouch!).

Momzer Bastard.

Naches Pride from your children (*Shlepping naches* from the *kinder*).

Nebbish Nerdy person.

Noodge To nudge/remind.

Nosh Snacking: something I love to do!

Noo So?

Nudnik An annoying person.

Over sholom Passed away.

Oy! An exclamation of surprise.

Oy veh! An exclamation of shock.

Parev Foods that contain no dairy products.

Pesach Passover.

Pish To urinate.

Pletzel Like a bagel without the hole and with poppy seeds and onion on top.

Polkeh Chicken drumstick.

Puppik Navel.

Putz Idiot.

Rachmones Pity.

Rebbe Rabbi.

Rebbitsen Rabbi's wife.

Rosh Hashana The Jewish New Year.

Saychel Common sense; using your head.

Seder The meal we eat first and second night of Passover.

Sefer Torah The scroll containing the five Books of Moses.

Sephadi A Jewish person from Spanish or Portuguese descent.

Shabbos The Sabbath.

Shadchen A matchmaker.

Shah Shut up!

Sha'koyach Congratulations.

Shamus No, not an Irish person – a detective.

Shayn Pretty.

Sheitel A wig worn by a married Orthodox woman.

Shiddach A possible marriage introduction.

Shiker Drunk.

Shissel A cooking pot.

Shivah A seven-day period of mourning.

Shlemiel An idiot.
Shlep To carry or to go a long way.
Shlepper Someone who carries all your heavy goods.
Shloch An untidy person.
Shlong The male organ (large).
Shluff To have a sleep.
Shmaltz Chicken fat. Or being too sentimental (no, I don't know why, either).
Shmatta Rags, cheap clothes. Can be said sarcastically when you are wearing your brand-new D&G.
Shmear To spread ('A *shmear* of cream cheese on my bagel, please.') Or to bribe.
Shmendrick An idiot.
Shmo An idiot.
Shmooze To coax with charming behaviour (JP's are very good at this).
Shmuck An idiot. Or a penis...
Shmutzy Looking unkempt or dirty.
Shnide To do something underhanded.
Shnorrer A mean person.
Shochet The ritual slaughter of animals for kosher meat.
Shofar A ram's horn blown in synagogues on the new year.
Sholom Peace.
Shpilkes When you are restless.
Shpritz A spray or squirt of something.
Shtetl A Jewish village in Eastern Europe.
Shtick Something that is funny.
Shtook In trouble.
Shtoom! Keep quiet! Can be said to children when noisy or when you want to keep something quiet.

Shtup A vulgar word for sexual intercourse; it also means tipped.
Shul Synagogue
Shvitz To be hot.
Shyster A thief or unscrupulous person.
Siddur A prayer book.
Simcha A joyous occasion. A common expression is 'Only on *simchas*'.
Smetana A type of sour cream.
Spiel A salesman's chat when he's talking to a client.

Tallis A Jewish prayer shawl (not a pashmina...).
Talmud Jewish law and tradition.
Tush Bottom.
Torah The Five Books of Moses.
Treif *See* page 18. Un-kosher – prawns, pork, etc.
Tsedrayt A mad person.
Tsures Troubles.
Tzaddik A righteous man.
Tzimmes (*see* page 98) A dish of cooked meat and vegetables, and sometimes with fruit.

Worsht (pronounced *vorsht*) Salami.

Yachna A gossip.
Yahrtzeit The anniversary of a death.
Yamulka Skullcap – same as *kapel*.
Yenta A female gossip.
Yiches Prestige.
Yiddishe Jewish.
Yiddishkeit Jewishness.
Yom kippur The Day of Attonement.
Yom tov A Jewish holiday.

Zaydeh Grandfather.
Zey gezunt Go in good health.

princess pointers

Every Princess knows that a little bit of advice can be invaluable (except when it comes from your mother-in-law). In the kitchen, having a few tips up your sleeve not only makes life easier, but it can also cut down on time, stop your hands smelling of onions and – most importantly – impress your mother-in-law. So here are a few that I find most handy.

1   If you place some kitchen paper folded double on top of your chicken soup, then lift it off, all the fat comes away with it.

2   If you're looking for *challah*, matzo meal, chicken-soup mix, dairy-free creamer or dairy-free margarine, just try your local kosher deli or the kosher section at your supermarket. Or better still, have a look at our website (*see* page 221).

3   When cutting fresh bread and cakes, first dip your knife in boiling water. You will get a lovely, clean slice.

4   A glass will not crack or break if you place a spoon in it first before pouring in hot liquid (Science Teacher, a.k.a. son aged 12, taught me that one).

5   When presenting food or arranging your flowers, always use odd numbers.

6   Prick sausages with a fork before cooking, as this prevents them from bursting.

7   When slicing an onion, don your rubber gloves to avoid smelly hands. Let the water run to avoid sliding mascara – or better still, buy frozen ready-sliced onions.

8   If your soup or stew is too salty, add a teaspoonful of sugar or a little grated raw potato and cook gently for a few minutes.

9   When chopping herbs, use a pair of kitchen scissors.

10   When peeling hard-boiled eggs, immerse them in cold water, then roll on kitchen paper to remove the shells.

11   If a guest offers to bring a dish, say 'YES!'

12   Always check your store cupboard for ingredients before baking.

13   Always, always turn the oven on *before* you start making a cake in order that it will reach the correct temperature.

14   Double up! If you're going to the trouble of making a cake, why not make two and freeze one? Then you're always prepared for those surprise guests.

15   Get the kids involved. This has many benefits. Firstly, it turns them on to cooking (future Princes and Princesses). Secondly, it makes them more adventurous when it comes to eating. Thirdly, they will be able to cook for you in your old age.

16   If you have great help, hang on to her (or even *him*). Don't forget to offer praise for doing a great job.

17   Always wash your hands before cooking, and after you have handled raw meat and eggs.

18   When using eggs, always crack them into a glass bowl one at a time, before adding them to any other ingredients. This helps avoid the shell going into the mixture, and you can also see if any are bad (or unkosher, if they contain blood spots).

19   Insert a sharp blade into a cake to see if it's done. If it comes out clean, your cake has baked.

20 To enable a cake to come out of the tin easily, make sure you grease (with butter) and flour it before pouring in the mixture – regardless of whether or not the recipe calls for it.

21 When removing a cake from a loose-bottomed tin, place the tin on a tin can and press down to lift off the sides.

22 Don't forget to use icing sugar – it's like fairy dust. Sprinkle over cakes and fruit for that little extra touch.

23 When using a loose-bottomed tin, wrap aluminium foil around the bottom to prevent the mixture from seeping out. This prevents your oven from getting dirty.

24 When you are at the checkout, don't be afraid to ask for a packer – or better still, shop online.

25 Check your sell-by dates. After all, you won't want to poison your family – so if in doubt, chuck it out.

26 Know your oven like a friend. They all vary, and some cook quicker than others, so keep this in mind when following recipes.

27 Don't forget: if you spend a long time in the car (and most JPs do this) always have water, tissues, coins for meters, Junior JP snacks, hand cream and paracetamol to hand.

28 When buying aluminium foil, baking parchment and cling film, always buy the jumbo size.

29 Wear your Jewish Princess apron when cooking to protect those couture creations.

30 Always wear your Jewish Princess oven gloves to protect your perfect manicure.

the jewish princess parting note

The Jewish Princess loves to party, but you don't need to go to a party in order to *parteee!* Every day can be a party. It's just how you decide to live your life.

Find your inner Princess and you will see that the little things in life – like eating with your friends and family, sitting down and having a nice cup of tea, laughing on the phone at some amusing tale, or treating yourself to a little something (I find handbags always cheer me up enormously); in other words, finding the things that make you happy (and often this means *making others* happy) – will show you that the Jewish Princess Champagne Theory is true.

'What's the Jewish Princess Champagne Theory?' I hear you ask.

Well, here it is:

'Is the Champagne Bottle of Life half-full or half-empty?'

For every Jewish Princess, the answer is: 'The Champagne Bottle of Life is ALWAYS half-full.'

But of course, she always orders another bottle to keep in reserve. (She can't help it. It's genetic.)

I know that the pace of this hectic world can sometimes make you feel like you're riding on a very fast train and hanging on for dear life; well, hang on in there, because eventually you *will* get to your destination. If you just try and arrive in style, however, no one will ever be the wiser.

Anyway, enough of Princess Philosophy.

What I really want to say is GOODBYE until the next time we meet, which I am sure will be very soon, as we live in a very small world. Of course, being a Jewish Princess I hate to say goodbye; I prefer to say *sholom.*

Peace to all my Princess People, whether you are Jewish, Christian, Catholic, Hindu, Muslim, Atheist... Let's try and find our common ground, so that we can come together and share our cultures and our recipes.

Let us sit down for dinner sometime and enjoy the simple things in life (like Bvlgari jewellery).

Now I really must go – I have a very important meeting. Yes, you guessed it: HAIRDRESSER!

I may be gone for a while, as I need my colour doing, so while I'm away, just remember the Three Ps:

* positive
* productive

And, of course:

* princesslike!

XXXXXXXXXXX

if all else fails…

…call the caterer!

www.thejewishprincess.com